Springer Theses

Recognizing Outstanding Ph.D. Research

Aims and Scope

The series "Springer Theses" brings together a selection of the very best Ph.D. theses from around the world and across the physical sciences. Nominated and endorsed by two recognized specialists, each published volume has been selected for its scientific excellence and the high impact of its contents for the pertinent field of research. For greater accessibility to non-specialists, the published versions include an extended introduction, as well as a foreword by the student's supervisor explaining the special relevance of the work for the field. As a whole, the series will provide a valuable resource both for newcomers to the research fields described, and for other scientists seeking detailed background information on special questions. Finally, it provides an accredited documentation of the valuable contributions made by today's younger generation of scientists.

Theses are accepted into the series by invited nomination only and must fulfill all of the following criteria

- They must be written in good English.
- The topic should fall within the confines of Chemistry, Physics, Earth Sciences, Engineering and related interdisciplinary fields such as Materials, Nanoscience, Chemical Engineering, Complex Systems and Biophysics.
- The work reported in the thesis must represent a significant scientific advance.
- If the thesis includes previously published material, permission to reproduce this must be gained from the respective copyright holder.
- They must have been examined and passed during the 12 months prior to nomination.
- Each thesis should include a foreword by the supervisor outlining the significance of its content.
- The theses should have a clearly defined structure including an introduction accessible to scientists not expert in that particular field.

More information about this series at http://www.springer.com/series/8790

Noah Mitchell

Geometric Control of Fracture and Topological Metamaterials

Doctoral Thesis accepted by
the University of Chicago, IL, USA

 Springer

Noah Mitchell
Kavli Institute for Theoretical Physics
University of California at Santa Barbara
Santa Barbara, CA, USA

ISSN 2190-5053   ISSN 2190-5061   (electronic)
Springer Theses
ISBN 978-3-030-36360-4   ISBN 978-3-030-36361-1   (eBook)
https://doi.org/10.1007/978-3-030-36361-1

This Springer imprint is published by the registered company Springer Nature Switzerland AG.
The registered company address is: Gewerbestrasse 11, 6330 Cham, Switzerland

In memory of Dr. Neville Eugene Prentice

Foreword

Concepts from geometry and topology can provide a powerful interpretative key in understanding the behavior of matter, from topological defects in ordered phases, to the self-assembly of material and biological matter driven by the geometry of their constituents. Conversely, these interpretative keys can be used to control and engineer the response of materials.

The work presented in this thesis explores this interplay in the study of propagating cracks and in the mechanical response of active metamaterials. Notably, the work is based on three distinct experimental systems, ranging in scale from centimeters to nanometers, as well as two classes of numerical methods. This reflects the best traditions of soft condensed matter in which a rich variety of classical many-body and material systems is called upon to reveal and put to the test unifying concepts in condensed matter. The freedom from a specific experimental platform allows soft condensed matter physicists a creative freedom to develop simple, experimentally accessible and often elegant experiments that reveal and isolate key aspects many-body physics.

The first part of this thesis demonstrates that geometry can control crack propagation. Draping thin sheets of material on curved surfaces changes the way cracks propagate within them, with their paths guided by the curvature of the surface. The work, whose experimental platforms are divided by six orders of magnitude in scale, presents conceptual, experimental, and numerical innovations.

From composite materials in aircraft to reconfigurable materials, biomimetic and biological materials, stretchable electronics, and sheets that snap, buckle, or rip on demand, this century marks the appearance of soft and composite materials into mainstream engineering. Correspondingly, these innovations and quests for versatility motivate us to revisit the exquisite intellectual tradition of fracture mechanics in rigid materials. Many questions remain ripe for exploration: from the simple failure mechanics of extremely soft materials to the question of how strongly curved and deformed materials modify crack behavior.

Chapters 2 and 3 of this thesis probe the question of how the Gaussian curvature of the substrate modifies the propagation of creeping cracks. Underlying this conceptual question are two non-local effects: cracks propagate in directions determined as much by their history as their local environment, and Gaussian curvature leads to non-local stress distributions in materials.

To isolate the effects of substrate curvature on crack propagation required identifying and characterizing a material whose resistance to fracture would be large enough so that a sheet can be strongly curved, while small enough that the stresses arising from conformation to substrate curvature are sufficient to drive and steer crack propagation. Chapter 2 of this thesis presents a new model material for this purpose, called "Rubber Glass" experiments based on this material adhered to 3D printed landscapes reveal a clear interaction between the curvature and the crack path.

To extend these observations beyond the experimental scales accessible, a numerical scheme was developed, based on the so-called phase field models of fracture mechanics. Fracture mechanics is a delicate business: stress fields diverge at the crack tip, making the problem singularly challenging to address theoretically. The physicist's approach to fracture mechanics has sought to pose the equations of motion for crack trajectories in a field theoretic context, with the aim of building a self-consistent formulation. Proceeding by analogy with phase transitions, phase field models have proven a successful numerical approach. Chapter 2 presents an extension of this effort to incorporate substrate curvature in a phase field models that can readily be implemented on a laptop computer to predict the motion of cracks in flat sheets conformed to curved surfaces.

Reducing the scale by six orders of magnitude, Chap. 3 explores how these notions of controlling failure via curved substrate topography generalize down to the length scales of nanoparticle sheets. In this context, additional forces, such as van der Waals forces, become important for the resulting failure energetics and crack morphologies. The lattice structure of the material also plays an important role: the material can not only rip, but also create dislocations in the lattice structure. Chapter 3 shows that substrate curvature tunes nanoparticle sheets between three regimes: gentle draping, azimuthal fracture, and radial folding. This work rationalizes the transitions between regimes through a phenomenological model, predicts stress patterns through spring network simulations, and explains the emergence of plastic deformation patterns in the ultra-thin nanoparticle sheets as they irreversibly adhere to lattices of spheres hundreds of nanometers in size.

The second half of the thesis focuses on the concept of topological order hidden in the mechanical response of materials. The notion that topology could play a role in the dynamics of matter has fascinated physicists since early studies of holonomy in mechanics, and gained popularity since the breakthroughs by Berry and later condensed matter theorists such as Thouless. In recent years, there has been an explosion of interest in the effects of hidden topological order in electronic systems. This has led to a re-examination of the role of topology in classical systems pioneered by Haldane.

Chapters 4–6 show that a collection of gyroscopes coupled by springs naturally possesses hidden topological order. Understanding the mechanical response of materials by modeling them as a collection of masses coupled by springs has been the mainstay of solid-state physics for decades. In this thesis, a seemingly simple ingredient is added: spin. This seemingly innocent twist on masses on springs brings topological phenomena to life. A collection of what are effectively spinning tops suspended from the ceiling show remarkable mechanical responses when poked. Instead of waves propagating throughout the structure as well as around the edges, a cohesive wave packet propagates around the edge of the material. What is even more remarkable is that if a notch is placed in the way of this wave packet, the wave packet simply avoids it! Ordinary waves in materials would scatter, being partially reflected and transmitted. This robustness is the most visible manifestation of the presence of underlying topological order.

What is more, this topological order, commonly associated with long-range spatial order of the underlying structure, depends only on the very *local* structure of the network. The simplicity of the approach allowed the construction of amorphous networks in any platform and was soon followed by work in electronic areas and is having a growing influence in the field.

The breadth and character of the work presented in this thesis as a whole delightfully reflects the creative process of science, whose most important and useful applications almost inevitably follow from the application of the scientific method to the curious pursuit of ideas and their relationship to reality.

Chicago, IL, USA   William Irvine
October 2019

Preface

Geometry and topology have emerged as powerful tools for understanding a wide range of phenomena in condensed matter physics. Often, geometric and topological constraints drive the order and dynamics of soft mechanical systems—systems in which material behavior is energetically accessible at room temperature in tabletop experiments. Here, we investigate the role of geometry and topology in shaping the mechanics of quasi-two-dimensional elastic materials.

In this thesis, we first present curvature as a geometric tool for guiding the behavior of cracks. When a flat elastic sheet conforms to a surface with Gaussian curvature, the geometry of the surface redistributes stresses in the sheet in a tunable fashion. Using this insight, we uncover how curvature can stimulate or suppress the growth of cracks and steer or arrest their propagation. We examine the mechanics of this scenario with and without pinning, in systems on both macroscopic and nanometric scales. Potential applications of the results range from stretchable electronics to functionalized crack patterns on the micron scale.

We then turn to discrete metamaterials for which network geometry and real-space lattice topology interact with the structure of elastic waves. In these mechanical materials, topological order in their excitation spectra translates to exotic behaviors at the materials' boundaries, such as chiral edge waves that are unusually robust to disorder. We uncover such topological behavior in a simple system composed of interacting gyroscopes and use this metamaterial to explore broken symmetries and tune through topological phase transitions. We then peel away a canonical ingredient for constructing topological insulators: the ordered underlying lattice. We find topological physics emerging from amorphous networks of gyroscopes and establish the basic building blocks for understanding topology in amorphous systems more generally. The results apply to a broad class of systems, from acoustic and mechanical structures to electronic and photonic materials. This work forges a path towards designing materials that transmit energy, sound, or light along their boundaries in a robust manner, even when the bulk of the material is messy or disordered.

Santa Barbara, CA, USA   Noah Mitchell

Acknowledgments

Foremost, I would like to thank Shannon Nicholson, for her patience, love, joy, enduring curiosity, and love of learning. Thanks to my brother, David Mitchell, for filling life with adventures and buttressing my interest in science. Thanks also to my parents for cultivating joy in undertaking arduous projects and to my extended family for embodying personal and scientific virtues.

I am incredibly grateful to William Irvine for support along many axes, including guiding the questions I ask, shaping my scientific writing and presentations, and encouraging me to explore a variety of topics. Vincenzo Vitelli, Heinrich Jaeger, Sid Nagel, and Tom Witten have each acted as strong mentors, both scientifically and personally. I am incredibly grateful to each of them. I am grateful to Paolo Privitera for welcoming me to the University of Chicago, as well as to Michael Levin for his insights. Paolo, Michael, Sid, and William have served as a wonderful thesis committee. Ari Turner has provided extensive insight into topological mechanics that have shaped much of this work; I value my time sharing in his creative and playful outlook. Evan Skillman, Kristy McQuinn, David Nitz, Amy Kolan, and David Dahl were phenomenal advisors during my undergraduate years.

The work in this thesis is the result of numerous collaborations, and no project has been entirely my own. Vinzenz Koning, Lisa Nash, Takumi Matsuzawa, Remington Carey, Daniel Hexner, and Stéphane Perrard have made collaborations a pleasure. Many other students and postdocs have contributed greatly to my scientific and personal development during my time at Chicago as well. Efi Efrati and Dustin Kleckner acted as guides in my initial plunge into soft matter and held my work accountable to a high standard. Thanks also to Vishal Soni and Martin Scheeler for keeping our office members grounded. I am thankful to the entirety of the Irvine Lab, especially Hridesh Kedia, Sofia Magkiriadou, Robert Morton, and Ephraim Bililign.

Alex Edelman, Anton Souslov, Manos Stamou, Michael Rauschenbach, Kieran Murphy, Ivo Peters, Irmgard Bischofberger, Corentin Coulais, Brent Perreault, Jonah Simpson, Thomas Videbaek, Nate Earnest, Ravi Niak, Joey Paulsen, Marc Miskin, Sri Iyer-Biswas, Michelle Driscoll, Nick Schade, Arvind Murugan, Clara del Junco, Van Bistro, and many others have infused color into my years as a graduate student.

Parts of This Thesis Have Been Published in the Following Journal Articles

- Noah P. Mitchell, Vinzenz Koning, Vincenzo Vitelli, and William T. M. Irvine. Fracture in sheets draped on curved surfaces. *Nature Materials*, 16(1):89–93, January 2017.
- Noah P. Mitchell, Remington L. Carey, Jelani Hannah, Yifan Wang, Maria Cortez Ruiz, Sean P. McBride, Xiao-Min Lin, and Heinrich M. Jaeger. Conforming nanoparticle sheets to surfaces with Gaussian curvature. *Soft Matter*, (14)9107–9117, October 2018. (Adapted with permission from The Royal Society of Chemistry)
- Noah P. Mitchell, Lisa M. Nash, and William T. M. Irvine. Realization of a topological phase transition in a gyroscopic lattice. *Physical Review B*, 97(10):100302, March 2018.
- Noah P. Mitchell, Lisa M. Nash, and William T. M. Irvine. Tunable band topology in gyroscopic lattices. *Physical Review B*, 98(17):174301, November 2018.
- Noah P. Mitchell, Lisa M. Nash, Daniel Hexner, Ari M. Turner, and William T. M.Irvine. Amorphous topological insulators constructed from random point sets. *Nature Physics*, 14(4):380–385, April 2018.

Contents

Chapter 1
Introduction

Geometry is not only a language to explain phenomena of the natural world, but also a tool to organize and trigger specific behaviors in material systems. As Jean le Rond D'Alembert wrote in 1752, "Geometry, which must obey Physics only when it meets with it, sometimes commands it" [1, 2]. In patterned liquid crystals [3–6], DNA lattices [7, 8], colloidal crystals [9–11], and classic models of phase transitions [12, 13], geometric constraints offer a mechanism to drive the order and dynamics of soft matter systems, both in and out of equilibrium. When curvature acts as the driving constraint on a two-dimensional material, that material's constituents may no longer tile their preferred local arrangement throughout curved space (Fig. 1.1). The material may respond elastically by stretching and compressing to accommodate its new geometry, or by forming defects such as dislocations [14] and disclinations [5, 10]. Might we similarly use curvature to guide the material failure of thin elastic materials conformed to corrugated surfaces?

Beyond assemblies of discrete matter in real space, it is no surprise that geometric constraints may also constrain the wave properties of a material; for example, the normal modes of an elastic plate must differ from those of a spherical elastic shell [15]. More subtle, however, is the recent realization that bands of phononic excitations themselves may also be constrained by nontrivial topology in momentum space [16–18]. While the normal modes of a 2D material may naturally live on a torus, a topological obstruction may exist that prevents continuously connecting the phases of the normal modes on that torus [19] (Fig. 1.2).

While these ideas can quickly become abstract and technical, we focus our attention here to simple, concrete realizations. This thesis presents two principal efforts. In the first effort (Chaps. 2 and 3), we explore the mechanics of cracks and plastic deformation in thin sheets draped onto surfaces with Gaussian curvature. The second effort (Chaps. 4–6) aims to understand the topological aspects of elastic waves in 2D metamaterials, using networks of gyroscopes as a model system. Before embarking, we first illustrate the how the ideas of curvature and topology are intertwined and how these, in turn, are linked to mechanics.

© Springer Nature Switzerland AG 2020
N. Mitchell, *Geometric Control of Fracture and Topological Metamaterials*,
Springer Theses, https://doi.org/10.1007/978-3-030-36361-1_1

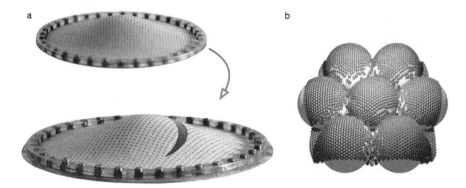

Fig. 1.1 Curvature can govern mechanical behavior. (**a**) Curvature of a rigid substrate stretches thin sheets, redistributing stresses. These stresses, in turn, can guide the paths of cracks. (**b**) The curvature of a corrugated substrate—here formed from a lattice of spheres—dictates the failure patterns of a nanoparticle sheet

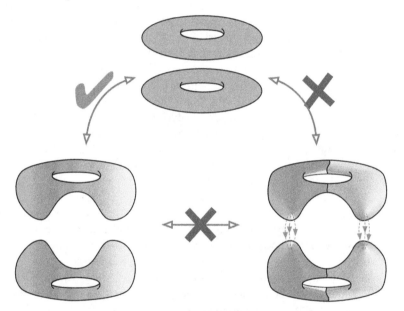

Fig. 1.2 Bands of normal modes in a dynamical system can exhibit nontrivial topology. In this illustration, each torus represents a band defined over the Brillouin zone of a lattice, and the color denotes the phase of the normal mode at each Brillouin zone wavevector, or each location on the torus. The phases of normal modes on the left can be continuously connected to the trivial case, in which the phases do not vary (top center). In the configuration on the right, there is a topological obstruction to a single-valued definition of the phases of normal modes in either band. The presence of phase defects—analogous to a Dirac monopole of positive (top band) and negative (bottom band) charge contained within each torus—presents a topological obstruction to covering either band without stitching together two phase descriptions

1.1 Curvature and Geometry

What does it mean for a surface to be curved? While we all have an intuitive notion of curvature, here we will require a more precise definition. A useful formulation is to measure the change of a surface's normal vector while moving tangent to that surface (Fig. 1.3a). Denoting the normal vector as $\mathbf{n}$ and vector tangent to the surface as $\mathbf{t}$, we may write the curvature along $\mathbf{t}$ as a tensor, $K_{\alpha\beta} = -\mathbf{t}_\beta \cdot \partial_\alpha \mathbf{n}$. At a given point on the surface, the matrix $K_{\alpha\beta}$ can be diagonalized to define eigenvectors (in this case, special tangent vectors) on the surface. The rate of change of the surface's normal vector along each of these directions can be described in terms of tangent circles with radii R_1 and R_2. We ascribe these radii a sign, determined by the side of the surface on which the tangent circle resides (Fig. 1.3b–d). Quite pleasantly, the invariants of the curvature tensor result in the mean curvature, H, and Gaussian curvature, G, as

$$2H = \mathrm{Tr}(K^\alpha_\beta) = \frac{1}{R_1} + \frac{1}{R_2} \tag{1.1}$$

$$G = \det(K^\alpha_\beta) = \frac{1}{R_1 R_2}. \tag{1.2}$$

Consider the difference between these two by bending this sheet of paper in this thesis. You can easily roll this page into a cylinder without stretching or compressing the sheet. Such a configuration has $H \neq 0$, and therefore we learn that, in the case of a cylinder, it requires very little energy to impart nonzero mean curvature to a thin sheet. Try as you may, however, it will be quite difficult to stretch this page onto a spherical surface without folding, crumpling, or ripping it apart.

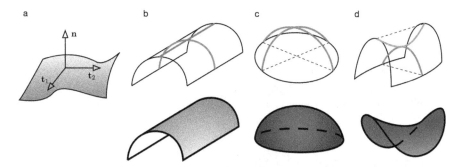

Fig. 1.3 Gaussian curvature is the product of principal curvatures at a given location on a surface. (**a**) A useful definition of curvature is the rate of change of the normal vector, $\mathbf{n}$, along a tangent vector of the surface, $\mathbf{t}_1$ or $\mathbf{t}_2$. (**b**) For a cylinder, one principal curvature vanishes, and therefore the Gaussian curvature is zero. (**c**) For a sphere, both curvatures have the same sign, so the Gaussian curvature is positive. (**d**) For a saddle, each curvature has opposite sign, so the Gaussian curvature is negative

In somewhat more general terms, we may deform a thin sheet into any developable surface—a surface with vanishing Gaussian curvature such as a plane, a cylinder, or a portion of a cone—without stretching or compressing the sheet. In each 3D geometry, the sheet is in some sense still flat. In this way, Gaussian curvature is *intrinsic* to the surface in a way that mean curvature is not: no matter how the sheet of paper is embedded in 3D space, it is still intrinsically 'flat' if it forms a developable surface.

Gaussian curvature is also a signed quantity. In a sphere, the tangent circles with radii $R_1 = R_2$ reside on the same side of the surface, in the interior (Fig. 1.3c). Thus, no matter if we denote these radii as positive or negative, their product is positive: $G > 0$. Similarly, any surface which curves in the same manner in all directions has positive Gaussian curvature at that point, regardless of whether we could call this point a peak or a valley. Negative Gaussian curvature surfaces, on the other hand, have one tangent circle on each side of the surface. Here too, the convention of which circle has a negative curvature is irrelevant for the sign of $G < 0$. Appendix A demonstrates an explicit construction of a surface of revolution with constant negative Gaussian curvature (a 'pseudosphere'), providing a more technical exposition to balance the qualitative discussion of this section.

What does the sign of Gaussian curvature signal? As a starting point, consider two airlines traveling due north from the equator, one from Quito and another from Nairobi. If the surface of the Earth were planar, they would not grow closer together as they fly north; their parallel paths would neither converge nor diverge. The two planes, however, grow nearer and nearer together due to the positive Gaussian curvature of the Earth's surface. In this way, $G \neq 0$ signals a deviation of initially parallel geodesics—which are, loosely speaking, the straightest lines possible on the surface. Denoting the distance between planes as ξ and their traveled distance s, the sign of the Gaussian curvature determines if these geodesics converge or diverge, according to the equation of geodesic deviation

$$\frac{d^2\xi}{ds^2} = -G\xi. \tag{1.3}$$

We will shortly examine how this sign is related to how a flat sheet must be stretched or compressed to adopt this shape, but first, let us consider the relationship between a surface's Gaussian curvature and its topology.

1.2   From Geometry to Topology

While the geometry of a surface describes how it is curved or embedded in space, the topology of a surface measures quantities which are invariant under continuous deformations of the surface, such as stretching or bending. Roughly speaking, while geometry measures, topology counts. As an example, consider the surface of a potato and the surface of a sphere. Though the Gaussian curvature of a potato varies along its surface, the potato and the sphere are topologically equivalent, in the sense that they both have the same *genus*: they both have no handles. Similarly, a coffee cup and a donut both have a genus of $g = 1$ since each has one handle, and they can likewise be smoothly deformed one into the other.

The geometry of a surface and the topology of that surface are not independent. Though curvature is a geometric feature that varies as a surface is stretched and deformed, the total integrated curvature of a surface is a topological invariant which cannot change unless the surface is ripped or punctured. This relationship is encoded in the celebrated Gauss-Bonnet Theorem

$$\int_{\mathcal{M}} G \, d^2\mathbf{x} + \int_{\partial M} k_g \, ds + \sum_i \theta_i = 2\pi \left[2(1 - g) - h\right]. \tag{1.4}$$

Here, G is the Gaussian curvature, k_g is the geodesic curvature along the boundaries ∂M of the manifold M (if there are any boundaries), the sum of θ's sums the external angles of any discontinuous turns in the boundaries, g is the genus of the surface counting the number of handles, and h is the number of boundaries or 'holes' of M. Any change to the curvature distribution of a smooth surface must satisfy the constraint of Eq. 1.4.

While Eq. 1.4 has been cast to describe smooth surfaces, a similar notion connects geometry to topology for discrete networks, such as the nanoparticle monolayers that we will discuss in Chap. 3. For a given surface, the Euler characteristic can be computed by sprinkling nodes on the surface and connecting nearby nodes via a triangulation (or some other polygonal tiling for which line segments connecting nearby nodes do not cross). Counting the number of vertices V, edges E, and faces F, one can then compute the Euler characteristic via

$$V - E + F = \chi, \tag{1.5}$$

where $\chi = 2(1 - g) - h$ is an integer topological invariant called the Euler characteristic. Just as Eq. 1.4 globally constrains changes in curvature or in the embedding of a smooth surface, so too Eq. 1.5 constrains changes in the local connectivity of nodes living on a surface.

The principles laid out above become physically manifest in a wide variety of contexts. Anyone who has taken a long flight knows how airlines' trajectories differ systematically from straight lines drawn on a 2D paper map. Other familiar examples are found in the design of domes, wherein a pentagon is often used as the capstone to allow much or all of the remaining surface to be tiled with hexagons. Figure 1.4b shows a Google Maps image of the Climatron in my hometown of St. Louis. In this structure, there is one disclination at the apex of the dome, surrounded entirely by hexagons. We may denote the disclination charge as concentrated at one location: $-\pi/3 \, \delta(\mathbf{x})$. Approximating the continuous portions of the boundary to be geodesics so that $k_g \approx 0$, the boundary of this dome must contain discontinuous

Fig. 1.4 Gaussian curvature constrains geodesic curves, tilings of space, and the existence of defects. (**a**) Two paths emanating north from the equator converge due to the positive Gaussian curvature of the sphere. (**b**) The Climatron in St. Louis, Missouri is a geodesic dome with a single, pentagonal disclination as its capstone and five kinks along its boundary. Topologically equivalent to a disk ($\chi = 1$), the dome compensates for the integrated Gaussian curvature via the disclination and the discontinuities along its boundary. Image courtesy of Google Maps (Imagery ©2018 Google, Map data ©Google) [20]. (**c**) The human papillomavirus (HPV) uses 12 pentagons to form a spherical viral capsid. Qutemol image of 1L0T courtesy of Dr. J.-Y. Sgro, UW-Madison (©2009 Jean-Yves Sgro) [21]

turns to create the remaining angular deficit necessary to balance the integrated Gaussian curvature:

$$\int_S \left(G - \frac{\pi}{3}\delta(\mathbf{x}) \right) \mathrm{d}S + \sum_i \theta_i = 2\pi\chi = 2\pi. \tag{1.6}$$

Here $\chi = 1$, since the surface has no holes ($g = 0$) and one boundary ($h = 1$). Closed spheres, such as soccer balls or certain viral capsids, have no boundary term, so that the disclination charges must cancel the integrated Gaussian curvature (Fig. 1.4c). For this reason, there are precisely 12 pentagons on any soccer ball that is otherwise tiled by hexagons, and 12 pentagonal elements in the human papillomavirus (HPV) capsid. The pentagonal elements are highlighted in purple in the capsid illustration in Fig. 1.4c, here constructed based on an atomic model [22] with the aid of Sybyl [23] and Qutemol [24].

These everyday examples relate mathematics to geometric aspects of physical space. However, they fall short of connecting curvature and topology to mechanics.

1.3 From Mathematics to Mechanics

Here, we begin to connect curvature and topology to the mechanics and elasticity of particular quasi-2D materials.

1.3.1 Curvature and Elasticity in Thin Sheets

Lay a thin sheet over a curved rigid substrate. If the sheet behaves inextensibly, like paper, then it must crumple, fold, or rip in order to accommodate the change in curvature. If the sheet is elastic, like rubber, then it can stretch and compress to adopt to the curved geometry. How will substrate curvature affect the stress distribution in the sheet? The answer will depend strongly on whether or not the sheet becomes pinned to the substrate as it conforms. For example, one way to conform an elastic sheet onto a curved surface is to glue every point on the undeformed sheet to the point on the surface directly below it in projection. If the sheet is allowed to relax, however, the displacements will rearrange: the azimuthal compression may relax, for instance, at the expense of stretching somewhat in the radial direction. Whereas the displacement in the pinned configuration depends on the details of how the sheet is conformed, such as which sites are pinned and in what order, the relaxed state is a unique configuration in the case without pinning.

Part I of this thesis considers how sheets draped on curved surfaces rip apart and plastically deform. We will first consider the case where no pinning occurs, using macroscopic rubber sheets as a model experimental system. Secondly, we will examine a specific case of fracture and deformation with strong pinning by draping nanoparticle membranes onto a lattice of rigid spheres. Here, we make a few introductory remarks linking Gaussian curvature to the elasticity of a sheet without the complications introduced by substrate pinning or material failure.

For thin sheets, the elastic free energy is dominated by stretching:

$$F = \frac{t}{2} \int \sigma_{ij} \varepsilon_{ij} \mathrm{d}A. \tag{1.7}$$

Here, the indices run over the two in-plane components, and t is the thickness of the sheet. How does the curvature arise in this expression? Here, we assume that the deflections of the sheet from its initial flat configuration are modest, so that the Föppl-von Kármán approximation holds; in particular, the surface normals remain approximately normal to the undeformed plane of the thin sheet. Under these assumptions, we can incorporate the corrugated profile of the substrate by including contributions in the height variations in the definitions of the 2D stress and strain directly. In particular, we couple the 2D displacements to the geometry of the substrate via $\sigma_{ij} = 2\mu\varepsilon_{ij} + \lambda\delta_{ij}\varepsilon_{kk}$ and $\varepsilon_{ij} = \frac{1}{2}\left[\partial_i u_j + \partial_j u_i + \partial_i h \partial_j h\right]$ for out-of-plane displacement $h(x, y)$. In this work, the effects from curvature come in through the gradients of h and we approximate $\mathrm{d}A = \mathrm{d}x\mathrm{d}y\sqrt{g} \approx \mathrm{d}x\mathrm{d}y$, which is justified for gentle deviations from flatness [14]. Minimization leads to the force balance equation

$$\partial_i \sigma_{ij} = 0, \tag{1.8}$$

which is automatically satisfied by introducing the Airy stress function, χ, such that

$$\sigma_{ij} = \varepsilon_{il}\varepsilon_{jk}\partial_l\partial_k\chi. \tag{1.9}$$

This yields the local differential equation

$$\frac{1}{Y}\nabla^4\chi = -G, \tag{1.10}$$

where G is the Gaussian curvature.

The meaning of Eq. 1.10 is that Gaussian curvature acts as a source for stress, in a manner analogous to charge sourcing an electrostatic field. In particular, the local Gaussian curvature acts as a charge-like source for the trace of the stress tensor, up to boundary terms. To see this, introduce the curvature potential $\Phi(\mathbf{x})$ such that

$$\nabla^2\Phi(\mathbf{x}) = -G(\mathbf{x}). \tag{1.11}$$

$\Phi(\mathbf{x})$ represents a scalar potential on the surface. Knowledge of $\Phi(\mathbf{x})$ then gives $\chi(\mathbf{x})$, and thus $\sigma(\mathbf{x})$, through

$$\frac{1}{Y}\nabla^2\chi = \Phi(\mathbf{x}) + H_R(\mathbf{x}) = \frac{\sigma_{kk}}{Y} \tag{1.12}$$

for some harmonic function $H_R(\mathbf{x})$, which is used to satisfy boundary conditions. If we lay aside boundary conditions for the moment, we see that the Gaussian curvature of the substrate acts as the source of the isotropic component of tension (σ_{kk}).

As we increase the geometric mismatch between the initial, flat state of our sheet and the final, curved state, stresses that build up in the sheet may trigger irreversible material failure. A pre-existing crack in the sheet (or one that grows from some other microscopic imperfection) will then interact with the curvature of the substrate through the stress distribution in the sheet. In a lattice-like material such as a nanoparticle sheet, defects such as dislocations may proliferate from the stresses caused by the geometric mismatch.

1.3.2   Berry Curvature, Chern Numbers, and Topological Mechanics

Beyond the elasticity of thin sheets, geometry and topology enter into the wave mechanics of 2D materials, creating new design principles for directing elastic waves.

The flow of energy and information is a central theme across condensed matter physics. The dispersion and energetics of phonons in particular has yielded numerous surprises, both in lattices and amorphous solids [25–27]. A major shift in this field came from studying not materials, but *metamaterials*—structures composed of

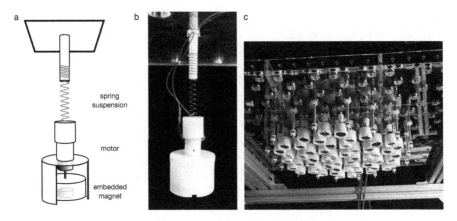

Fig. 1.5 The building block of the metamaterials studied here is a gyroscope hanging from a plate. The gyroscopes hanging from a plate are assembled into a 2D lattice or network and interact via magnetic repulsion

patterned elements [28, 29]. With this additional layer of complexity comes the ability to design structures from the bottom up. The ability to design the microstructural components has led to the fabrication of auxetics [30, 31], materials with negative index of refraction [32–35], and passive topological metamaterials [36, 37]. In patterning the microstructural components, degrees of freedom can be introduced or removed in the building blocks directly. Compounding this freedom with the ability to introduce activity in each component results in a broad palette for accessing exotic material properties. This realization has led to the design of a variety of phononic metamaterials, which are able to acoustically cloak inclusions [38], exhibit non-reciprocal wave properties [39, 40], and exhibit topologically-protected chiral waves on their boundaries [17, 41].

In Part II of this thesis, we focus on materials built from spinning gyroscopes. We couple the gyroscopes elastically via springs or magnetic interactions and pattern them in a 2D plane. Figure 1.5a, b show the principal ingredient: the gyroscope hanging from a rigid plate. Because the inertial response of this object is dominated by the angular momentum along the axis of rotation, an applied force reorients its displacement. If the gyroscope spins sufficiently rapidly that nutation is negligible, the free tip of a gyroscope moves with a velocity proportional to the torque acting about the pivot point, $\vec{\tau}$. This response is captured by Newton's second law in the form

$$\vec{\tau} \approx I\omega_0 \partial_t \hat{n} = \vec{\ell}_f \times \vec{F}, \tag{1.13}$$

where $\vec{\ell}_f$ is the vector from the pivot point to the point acted upon by force $\vec{F}$, I is the principal moment of inertia, ω_0 is the spinning speed of the gyroscope, and $\hat{n}$ is the unit vector pointing from the pivot to the center of mass.

For small displacements, we may write the displacement from the equilibrium position in the plane as $\psi = x + iy$. The equation of motion for a single gyroscope under the influence of gravity becomes

$$i\partial_t \psi = \frac{mg\ell_{cm}}{I\omega_0}\psi. \tag{1.14}$$

Two features of this equation stand out. First, the motion is perpendicular to the applied force, as seen from the presence of the imaginary factor, i. Second, the dynamics are first order in time, meaning that an applied torque results in a gyroscope's velocity. The similarity between Eq. 1.14 and the Schrödinger equation for a quantum particle encourages us to explore the topological mechanics of the phonons in gyroscopic networks.

As we will discover, networks of gyroscopes ubiquitously support topologically nontrivial band structure. Band gaps—ranges of frequencies in which no normal modes of the structure reside—are endowed with unidirectional chiral edge modes in these systems. Importantly, these edge modes are protected by the topological information encoded in the bands. Let us build up towards understanding what this means using curvature as our starting point.

Berry curvature is the direct analog of Gaussian curvature in a more abstract space: the space of normal modes. Just like Gaussian curvature, Berry curvature can be expressed as the curl of a connection, and intuitions of parallel transport on a surface embedded in three dimensions carry over to accumulating phase in the state of a system by moving through the configuration space of the system.

As a concrete illustration, consider an collection of interacting gyroscopes which oscillate in an eigenstate of the structure with eigenfrequency, ω. It will become useful to denote the evolution of a single gyroscope in terms of its right and left-circularly polarized components:

$$\psi_i = \psi_i^R e^{-i\omega t} + \overline{\psi}_i^L e^{i\omega t}, \tag{1.15}$$

where $\overline{\psi}_i^L$ is the complex conjugate of ψ_i^L. Denoting the collection of N gyroscope displacements as a list

$$|\boldsymbol{\psi}\rangle = (\psi_1^R, ..., \psi_N^R, \psi_1^L, ..., \psi_N^L), \tag{1.16}$$

the evolution in time of the eigenstate indexed by its wavevector $\mathbf{k}_0$ is then

$$|\boldsymbol{\psi}_n(\mathbf{k}_0, t)\rangle = e^{-i\omega t}|\boldsymbol{\psi}(\mathbf{k}_0, t = 0)\rangle. \tag{1.17}$$

Now, introduce an adiabatic force acting on the eigenstate that changes the energy and spatial wavevector $\mathbf{k}$ of the oscillating state. Sufficiently slow variations will keep the eigenstate $|\boldsymbol{\psi}(\mathbf{k}_0, t = 0)\rangle$ in an instantaneous eigenstate $|\boldsymbol{\psi}(\mathbf{k}(t), t)\rangle$. The later state is then given by

$$|\psi_n(\mathbf{k}, t)\rangle = e^{i\gamma_n(t)} e^{-i\int_0^t dt'\omega(\mathbf{k}(t'))}|\psi(t = 0)\rangle. \tag{1.18}$$

While the second exponential is simply the instantaneous eigenstate evolution, the first exponential term is geometric, with γ being the Berry phase. If we consider a cyclic evolution around a closed path of a state evolving according to Eq. 1.15, the Berry phase depends only on the path taken:

$$\gamma_n = i \oint_C d\mathbf{k} \langle \psi | \nabla_\mathbf{k} | \psi \rangle. \tag{1.19}$$

Just as parallel transport is determined by Gaussian curvature, here the Berry phase is determined by the Berry curvature

$$\vec{\Omega}_n = \nabla_\mathbf{k} \times \vec{A}_n(\mathbf{k}), \tag{1.20}$$

where $\vec{A}_n(\mathbf{k}) \equiv i\langle \psi | \nabla_\mathbf{k} | \psi \rangle$ is the so-called 'Berry connection' at wavevector $\mathbf{k}$. Integrating the Berry curvature over a patch of momentum space and using Stokes' theorem shows us that the Berry phase is determined by the enclosed Berry curvature:

$$\gamma_n = \oint_C d\boldsymbol{\ell} \cdot \boldsymbol{A}_n(\mathbf{k}) = \int_S d\mathbf{S} \cdot \boldsymbol{\Omega}_n(\mathbf{k}), \tag{1.21}$$

where $\mathbf{S}$ is a unit vector at $\mathbf{k}$ normal to the band.

We have seen that the Berry curvature of the normal modes determine a phase shift in the gyroscopes' displacement under adiabatic transformations about momentum space. In a lattice, momentum space can be viewed as a tiling of the Brillouin zone, just as the real space lattice configuration is a tiling of the unit cell. This means that the Brillouin zone is, topologically, a torus, and integrating the Berry curvature over its surface provides information about the connection of normal modes in each band.

Figure 1.6 shows two examples of a band colored by its Berry curvature. In Fig. 1.6a, the Berry phase accumulated around path C_1 is some nonzero number, but the phase accumulated around path C_2 is zero: the curvature contributions cancel. Likewise, the total curvature integrated over the whole band vanishes. In Fig. 1.6b, however, the contributions add, and the Berry phase accumulated around a path encircling the entire Brillouin zone is a nonzero multiple of 2π:

$$\oint_{C_2} d\boldsymbol{\ell} \cdot \boldsymbol{A}(\mathbf{k}) = \int_{S_2} d\mathbf{S} \cdot \boldsymbol{\Omega}(\mathbf{k}) = 2\pi\, C. \tag{1.22}$$

The integer C is the Chern number of the band.

It is not apparent at first how the Chern number is related to the waves of our material, beyond describing a topological aspect of the connection between normal modes. A powerful answer was provided by Thouless, Kohmoto, Nightingale, and

a   b

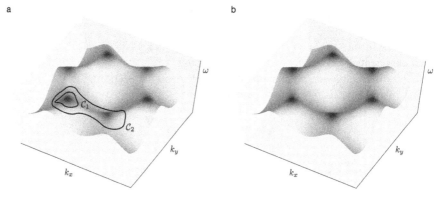

Fig. 1.6 Analogously to parallel transport on a curved surface, the Berry curvature indicates the phase picked up by adiabatic evolution along a path encircling some patch of a band of normal modes. (**a**) The phase accumulated by traversing $\mathcal{C}_1$ is nonzero, while the phase accumulated via $\mathcal{C}_2$ vanishes. (**b**) For a band with Chern number of $C = 1$, the integrated Berry curvature sums to 2π

de Nijs for the analogous problem of electronic Chern insulators: a nonzero Chern number signals the existence of chiral modes that live on the boundary of a system and are immune to back-scattering [42].

One intuitive explanation for the existence of these modes is that the bulk supports topologically nontrivial behavior described by a nonzero Chern number, while the absence of material beyond the boundary is equivalent to a trivial insulator. At the interface, something discontinuous must happen to the band structure. In the simplest scenario, the gap closes, leading to conducting surface waves on the boundary alone, and the Chern number denotes the number of right moving modes minus the number of left-moving chiral modes, as illustrated schematically in Fig. 1.7. The power of these chiral waves' topological origin becomes truly apparent only in the presence of disorder. Disorder could take the form of random variations in the masses or spinning speeds of gyroscopes, inclusions or voids in the material, or jagged features like corners at the material's boundary. Typically, in elastic systems, propagating waves will scatter off such disorder, resulting in interference and energy lost to reflections. Topological protection ensures that, in the absence of available counter-propagating states, an edge wave experiences perfect transmission, passing around inclusions or voids and readily changing direction along jagged boundaries (Fig. 1.7).

As we discover in Chap. 6, our understanding of Chern number generalizes beyond periodic lattices, into networks with amorphous (or 'glassy') spatial structure. In that exposition, our description of band topology is no longer built on Berry curvature directly, relying instead on the algebraic properties of projection operators.

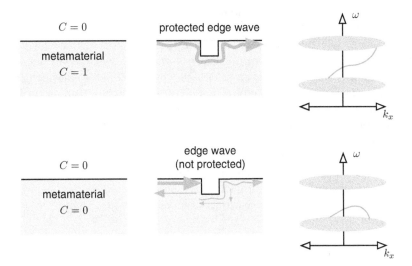

Fig. 1.7 The Chern number counts the number of modes occupying a band gap. In a system with a single gap and a single boundary, the number of right-moving modes—i.e. sets of possible eigenstates with positive group velocity—minus the number of left-moving modes is determined by the Chern number of a band. Nonzero Chern numbers give rise to topologically protected chiral edge modes, which are robust against back-scattering

1.4   Scope of This Book

As we have previewed, geometry and topology have emerged as powerful tools for understanding a wide range of phenomena in condensed matter physics. Often, geometric and topological constraints drive the order and dynamics of soft mechanical systems—systems in which material behavior is energetically accessible at room temperature in tabletop experiments. Here, we investigate the role of geometry and topology in the mechanics of quasi-two-dimensional elastic materials.

In Part I, which is based on references [43, 44], we examine how substrate geometry controls material failure. Chapter 2 uncovers the interaction between cracks in thin elastic sheets and Gaussian curvature of a rigid substrate. In particular, we study the energetics and the paths of cracks in flat elastic sheets conformed to surfaces with Gaussian curvature in the absence of pinning. Using rubber sheets as a model experimental system, the analysis of Chap. 2 applies generally to all materials that are thin, isotropic, linearly elastic, and brittle. Chapter 3, on the other hand, focuses on a particular material: nanoparticle monolayer sheets. These ultra-thin sheets are inorganic-organic hybrid materials are known for their controllable optical and electrical properties, mechanical flexibility, and remarkable strength. This material adds two ingredients to our study of material failure on curved surfaces: the discrete lattice structure of the material and the addition of van der Waals forces pinning the sheet to the substrate. Both ingredients play a strong role in determining the fracture morphology and deformation in nanoparticle sheets stamped onto curved surfaces.

In Part II, which is based on references [45–47], we examine topological mechanics of gyroscopic networks. Chapter 4 describes the experimental system used and introduces a mechanism to drive a topological phase transition in real time. Chapter 5 reports the study of a variety of gyroscopic lattices, including geometries for which the topological band structure can be tuned through bond-length-preserving deformations. Looking beyond periodic metamaterials, Chap. 6 reports the discovery of topologically-protected chiral edge waves in amorphous and aperiodic networks. We use a generalized notion of Chern number to explain the topological protection and apply it to our mechanical system. Furthermore, we demonstrate the generality of our findings and predict the existence of glassy topological insulators in electronic materials, photonic materials, acoustic resonators, and beyond.

Part I
Gaussian Curvature as a Guide for Material Failure

Chapter 2
Fracture in Sheets Draped on Curved Surfaces

Conforming materials to rigid substrates with Gaussian curvature—positive for spheres and negative for saddles—has proven a versatile tool to guide the self-assembly of defects such as scars, pleats [2, 9, 10, 14, 48], folds, blisters [49, 50], and liquid crystal ripples [3]. Here, we show how curvature can likewise be used to control material failure and guide the paths of cracks. In our experiments, and unlike in previous studies on cracked plates and shells [51–53], we constrained flat elastic sheets to adopt *fixed* curvature profiles. This constraint provides a geometric tool for controlling fracture behavior: curvature can stimulate or suppress the growth of cracks and steer or arrest their propagation. A simple analytical model captures crack behavior at the onset of propagation, while a two-dimensional phase-field model with an added curvature term successfully captures the crack's path. Because the curvature-induced stresses are independent of material parameters for isotropic, brittle media, our results apply across scales [54, 55]. This chapter is adapted from [43] with permission.

2.1 Gaussian Curvature as a Tool

Geometry on curved surfaces defies intuition: 'parallel' lines diverge or converge as a consequence of curvature. As a result, when a thin material conforms to such a surface, stretching and compression are inevitable [2]. As stresses build up, the material can then respond by forming structures such as wrinkles or dislocations, which are themselves of geometric origin. This interplay between curvature and structural response can result in universal behavior, independent of material parameters [9, 10, 14, 48, 50].

A markedly different material response is to break via propagating cracks. While the use of curvature to control the morphology of wrinkles and defects in materials has been recently explored [9, 10, 50], here we investigate the control of cracks by

© Springer Nature Switzerland AG 2020
N. Mitchell, *Geometric Control of Fracture and Topological Metamaterials*,
Springer Theses, https://doi.org/10.1007/978-3-030-36361-1_2

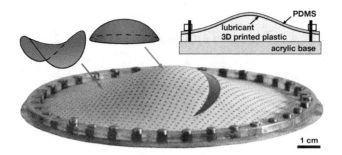

Fig. 2.1 Gaussian curvature—positive for caps and negative for saddles—governs the behavior of cracks. In the experimental setup, an initially flat PDMS sheet conforms to a curved 3D printed surface. A small incision nucleates the crack

tuning the geometry of a rigid substrate. Can we design the underlying curvature of a substrate to steer paths of cracks in a material draped on that surface, thereby protecting certain regions?

To probe the effect of curvature on cracks, we conform flat PDMS sheets (Smooth-On Rubber Glass II) to 3D-printed substrates (Fig. 2.1). A lubricant ensures that the sheet conforms to the substrate while moving freely along the surface. We consider various geometries having positive and negative Gaussian curvature in both localized and distributed regions, including spherical caps, saddles, cones and bumps. To begin, we focus on the bump as a model surface, as it is a common geometry containing regions of both positive and negative curvature. We seed a crack by cutting a slit in the sheet, with a position and orientation of choice. By successive cuts, we increase the slit's length until it exceeds a critical length, known as the Griffith length [56, 57], and propagates freely.

2.2   Fracture Onset: Griffith Lengths and Crack Kinking

The Griffith length of a crack in a flat sheet is nearly independent of position and orientation. On our curved geometry, we find that this is not so. On the top of the bump, a shorter slit is necessary to produce a running crack, and on the outskirts of the bump (where the Gaussian curvature is negative), the behavior depends strongly on the orientation of the seed crack: fracture initiation is suppressed for radial cracks, while the Griffith length for azimuthal cracks approaches that of the flat sheet (Fig. 2.2b). Thus curvature can both stimulate and suppress fracture initiation, depending on the position and orientation of the seed crack relative to the curvature distribution.

To relate these findings to the curvature distribution, we consider the stresses induced by curvature and their interaction with the crack tip. Stresses generated in the bulk of a material become concentrated near a crack tip. In turn, a crack extends when the intensity of stress concentration exceeds a material-dependent,

Fig. 2.2 Curvature stimulates or suppresses fracture initiation. (**a**) Gaussian curvature and curvature potential distributions for a bump with height profile $h(\rho) = \alpha x_0 \exp(-\rho^2/2x_0^2)$. (**b**) While the Griffith length for a crack in a flat sheet (dashed line) is nearly constant, curvature modulates the critical length of a seed crack. All samples shown had a 12 cm diameter ($2R$), an aspect ratio $\alpha = 1/\sqrt{2}$, bump width $x_0 = R/2.35$, and constant radial displacement $u_\rho/R = 0.012$

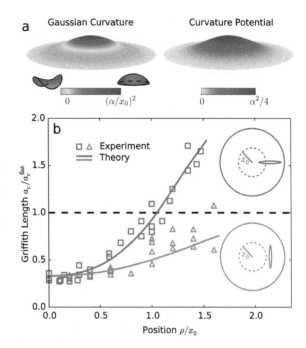

critical value [56, 58]. Expressed mathematically, in the coordinates of the crack tip (r, θ), the stress in the vicinity of the tip takes the form

$$\sigma_{ij} = \frac{K_I}{\sqrt{2\pi r}} f_{ij}^I(\theta) + \frac{K_{II}}{\sqrt{2\pi r}} f_{ij}^{II}(\theta), \qquad (2.1)$$

where $f_{ij}^{I,II}$ are universal angular functions [58]. The factors K_I and K_{II} measure the intensity of tensile and shear stress concentration at the crack tip, respectively, and are known as stress intensity factors. Thus, the Griffith length, a_c, is the length of the crack at which the intensity of stress concentration reaches the critical value, K_c. In curved plates or sheets, the near-tip stress fields display the same singular behavior as in Eq. 2.1 [59], but the values of the stress intensity factors are governed by curvature.

Curving a flat sheet involves locally stretching and compressing the material by certain amounts at each point. According to the rules of differential geometry, the amount of stretching, controlled by the field Φ, is determined by an equation identical to the Poisson equation of electrostatics [60], with the Gaussian curvature, G, playing the role of a continuous charge distribution [2, 14]:

$$\nabla^2 \Phi(\mathbf{x}) = -G(\mathbf{x}). \qquad (2.2)$$

As the sheet equilibrates, its elasticity tends to oppose this mechanical constraint, giving rise to stress. The isotropic stress from curvature is then related to the potential via $\sigma_{kk}^G = E\Phi$, where E is Young's modulus, and the stress components are determined by integrals of the potential and boundary conditions. In particular, using the definitions for the boundary contribution H_R defined in Eq. 1.12 of the introduction, the equations

$$\sigma_{\rho\rho}(\rho) = \frac{E}{\rho^2} \int_0^\rho \rho' \left[\Phi(\rho') + H_R \right] d\rho' \tag{2.3}$$

and force balance in polar coordinates

$$\sigma_{\phi\phi} = \sigma_{\rho\rho} + \rho \partial_\rho \sigma_{\rho\rho} \tag{2.4}$$

together fully specify our system.

Our study rests on a general geometric principle: positive (negative) curvature promotes local stretching (compression) of an elastic sheet, leading to the enhancement (suppression) of crack initiation. Variations in the potential Φ steer the crack path, with the form of Φ determined nonlocally from the curvature distribution.

For the bump, the curvature potential, Φ, is large on the cap, where curvature is positive, and decays to zero as the negative curvature ring screens the cap (Fig. 2.2a). Since $Y\Phi$ is the isotropic stress, crack growth is stimulated where the potential is greatest—on the cap of the bump, resulting in a small Griffith length there (Fig. 2.2b). Moving away from the cap, the potential decays, producing a stress asymmetry. This results in longer Griffith lengths with strong orientation dependence on the outskirts of the bump. Figure 2.2b shows the theoretical results overlying the experimental data, with no fitting parameters. We find that this minimal model is sufficient to capture the phenomenology of our system at the onset of fracture and provides correct qualitative predictions for longer cracks, even in the absence of symmetry.

2.2.1 Griffith Length for a Small Crack

For sufficiently small cracks, we may compute the Griffith length analytically for symmetric curvature distributions analytically. Given knowledge of the stress field of an uncracked sheet on a curved surface, we can then compute the stress intensity factors (SIFs), denoted K_I and K_{II}, for a crack on that curved surface, following our model. These quantities measure the intensity of tensile and shear stress concentration at the crack tip. The SIFs for each seed crack position and orientation follow from integrating the infinite-plane Westergaard solution over the crack length [61]:

$$K_{I,II} = \frac{1}{\sqrt{\pi a}} \int_{-a}^{a} d\xi \sqrt{\frac{a+\xi}{a-\xi}} \tilde{\sigma}_{I,II}(\xi, 0) \tag{2.5}$$

where $\tilde{\sigma}_I(x, y) = \tilde{\sigma}_{yy}(x, y)$ and $\tilde{\sigma}_{II}(x, y) = \tilde{\sigma}_{xy}(x, y)$ are the tensile and shear stresses, respectively, in the crack's local xy coordinate system. The tilde distinguishes $\tilde{\sigma}_{ij}(x, y)$ from $\sigma_{ij}(\rho, \phi)$, which is a function of the material coordinate system rather than the crack coordinate system and is therefore a different function of its arguments, despite being the same physical quantity. The stresses, $\tilde{\sigma}_{I,II}$, in Eq. 2.5 include both geometric frustration and additional boundary loading.

For cracks that are small compared to the length scale over which the stress fields vary $(a|\partial_x \tilde{\sigma}_{iy}| \ll \tilde{\sigma}_{iy})$,

$$K_I = \sqrt{\pi a} \left(\sigma_{\rho\rho}(\rho^*) \sin^2 \beta + \sigma_{\phi\phi}(\rho^*) \cos^2 \beta \right), \tag{2.6}$$

$$K_{II} = \sqrt{\pi a} \left(\sigma_{\phi\phi}(\rho^*) - \sigma_{\rho\rho}(\rho^*) \right) \sin \beta \cos \beta, \tag{2.7}$$

where the inclination angle β is the angle of the seed crack with respect to the radial direction. The Griffith length then takes the form

$$a_c = \frac{4K_c^2}{\pi Y^2 \left[2\Phi F(\bar{\theta}, \beta) \cos \beta - \Omega F(\bar{\theta}, 2\beta) \right]^2}, \tag{2.8}$$

where $F(\theta, \beta) = f_{\theta\theta}^I(\theta) \cos \beta + f_{\theta\theta}^{II}(\theta) \sin \beta$. To include boundary effects in the above, take $\Phi(\mathbf{x}) \to \Phi(\mathbf{x}) + H_R(\mathbf{x})$. To gain intuition from Eq. 2.8, note that the curvature potential measures the local isotropic compression: $\Phi(\mathbf{x}) = [\tilde{\sigma}_{xx}(\mathbf{x}) + \tilde{\sigma}_{yy}(\mathbf{x})]/Y$. This implies that crack growth tends to be suppressed in regions where $\Phi < 0$ and stimulated where $\Phi > 0$. A local stress asymmetry, however, can play an important role in attenuating this generalization. For instance, the orientation dependence of the Griffith length shown in Fig. 2.2b of the main text shows the importance of stress asymmetry in the determination of the Griffith length. Qualitatively, a curvature potential which increases with radial distance (a potential 'well') preferentially stimulates the growth of cracks which are oriented along the radial direction, so that the Griffith length of a radial crack in a potential well is smaller than that of an azimuthal crack centered the same distance ρ^* from the minimum of Φ. Conversely, potentials which decrease with distance from the center preferentially stimulate the growth of cracks oriented along the azimuthal direction.

2.2.2   Crack Kinking

Curvature not only governs the critical length for fracture initiation, but also the direction of a crack's propagation. For cracks inclined with respect to the bump, the cracks change direction as they begin to propagate, kinking at the onset of crack growth and curving around the bump, as shown in Fig. 2.3a. Cracks kink and curve towards the azimuthal direction because a decaying curvature potential, $\Phi(\rho)$,

Fig. 2.3 Kinking and
curving of crack paths in
sheets conformed to a bump.
(**a–b**) Crack paths kink and
curve around a bump. (**c–d**)
Phase-field simulations of
cracks on a bump, colored by
the phase-modulated energy
density so that broken regions
are darkened. (**e–f**) The
phase-field crack path
predictions (black solid
curves) overlie the
experimental paths (colored
curves). *(Inset)* Introducing a
time delay that matches
experiment for the right crack
tip's propagation eliminates
the discrepancy far from the
bump. (**g**) Analytical
prediction (solid black curve)
of the kink angle, θ_k, overlies
experimental results. (**h**)
Analytical crack path
predictions overlie
simulations for free (constant
stress) boundary conditions.
All experiments and
simulations have aspect ratio
$\alpha = 1/\sqrt{2}$ and bump width
$x_0 = R/2.35$, including the
free boundary condition
simulations

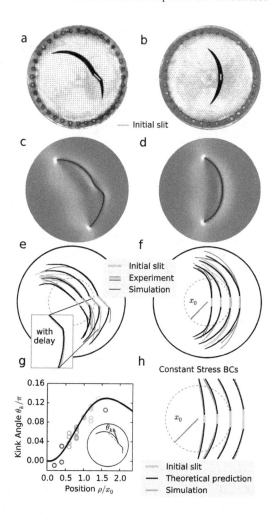

creates a local stress asymmetry: $\sigma_{\phi\phi}^G < \sigma_{\rho\rho}^G$. As a result, the crack relieves more
elastic energy by deflecting towards the azimuthal direction. Analytical prediction
of the kink angle, θ_k, is made by selecting the direction of maximum hoop stress
asymptotically near the crack tip:

$$\theta_k = 2 \arctan \left(\frac{2\eta}{1 + \sqrt{1 + 8\eta^2}} \right), \tag{2.9}$$

where $\eta \equiv K_{II}/K_I$. Figure 2.3g shows calculations that include the finite size of
cracks, yielding good agreement with experiment.

If we again consider small cracks, we can find expressions for the kink angle
analytically. For any rotationally symmetric curvature distribution $G(\rho)$, we can

invoke Eq. 2.3 and $Y\Phi = (\sigma_{\rho\rho} + \sigma_{\phi\phi})$ (dropping boundary effects for now) to find the ratio of SIFs of a small crack centered at ρ^*:

$$\eta = \frac{(\Phi - \Omega)\sin 2\beta}{\Phi + (\Phi - \Omega)\cos 2\beta},\tag{2.10}$$

where $\Omega(\rho^*) \equiv \frac{2}{\rho^{*2}}\int_0^{\rho^*}\rho'\Phi(\rho')d\rho'$ is the average value of the curvature potential in the region enclosed by the circle of radius ρ^*. Thus, the quantity $\Phi - \Omega$ appearing in Eq. 2.10 is the difference between the local value of the potential $\Phi(\rho^*)$ and the value of the potential averaged from the center to the location of the crack, and this quantity can be readily identified as the local stress asymmetry

$$\Phi - \Omega = \left(\sigma_{\phi\phi} - \sigma_{\rho\rho}\right)/Y.\tag{2.11}$$

For the crack to propagate, the tractions along the crack faces must be positive. As a consequence, the sign of this stress asymmetry determines whether the crack kinks towards the radial or azimuthal direction.

For a crack in a potential 'well' (where Φ increases with radial distance), the crack kinks toward the radial direction (with respect to the center of the well). For a crack in a potential 'dome' or 'peak' (where Φ decreases with radial distance) the crack kinks toward the azimuthal direction.

2.3 Crack Trajectories

Having captured crack behavior at the onset of propagation, we now turn to understanding the path cracks take in sheets draped on curved surfaces. We find that a purely analytical model is sufficient to capture the long-time behavior of the crack if the stress is fixed at the boundary (see Fig. 2.3h). We then introduce to a more robust phase field model approach which allows for arbitrary boundary conditions, at the cost of increasing computational complexity (Fig. 2.3c–f).

2.3.1 Perturbation Theory Prediction of Crack Paths

The perturbation theory approach of Cotterell & Rice [62] allows for the analytic prediction of the crack trajectory when the deflection from a straight path is small, as depicted in Figs. 2.4 and 2.5. Here we apply this formalism to the case of a flat sheet conformed to a curved substrate. Beginning with a straight seed crack, we compute curved path by iteratively updating the SIFs using the initial stress distribution of the curved sheet (see below), calculating the kink angle using Eq. 2.9 and extending the crack in the direction specified by θ_k by a small increment.

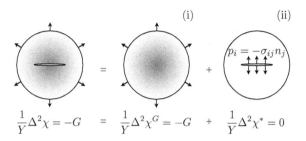

Fig. 2.4 Crack energetics and paths can be predicted analytically by a linear decomposing of the elastic situation. Linear decomposition of a frustrated, cracked linear elastic sheet into (i) a curved sheet with no crack and stress P at the boundary, and (ii) a flat sheet with tractions, **p**, on the crack faces

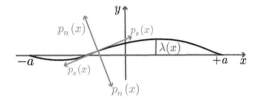

Fig. 2.5 Perturbation theory approach predicts crack paths for weak deviations from straight paths. A slightly curved crack with crack tips at $x = -a$ and $x = a$ has normal and shear tractions $p_{n,s}$ and deviation $\lambda(x)$

When contributions to the Muskehishvili potentials [63] are kept to linear order in the deflection of a crack from a straight path, $\lambda(x)$, the perturbation theory approach results in an expression for the stress intensity factors [62]:

$$K_I - iK_{II} = \frac{1}{\sqrt{\pi a}} \int_{-a}^{a} (q_I - iq_{II}) \sqrt{\frac{a+t}{a-t}} dt, \tag{2.12}$$

with

$$q_I = p_n - \frac{3}{2}\omega p_s + \lambda p_s' + 2\lambda' p_s \tag{2.13}$$

$$q_{II} = p_s + \lambda p_s' + \frac{1}{2}\omega p_n, \tag{2.14}$$

where $p_n(x)$ and $p_s(x)$ denote the normal and shear tractions, primes denote derivatives, and $\omega = \lambda'(a)$ is the slope of the crack at the tip. Note that Eq. 2.12 reduces to Eq. 2.5 when the crack is straight. In the presence of curvature, we compute the tractions from the curved sheet without a crack, as depicted in Fig. 2.4. Extending this analysis to include higher order terms of the Williams expansion for the stress provides an avenue for further investigation [64].

In order to be analytically tractable, this approach assumes the interaction of the crack with the boundary of the sample is negligible. We thus perform the perturbation theory approach for constant stress boundary conditions, fixing the radial stress at the boundary $\sigma_{\rho\rho}(\rho = R) = 0$, as shown in Fig. 2.3h. These paths closely match simulations using constant stress boundary conditions, with the same geometry and sample size as the experiments and simulations in Fig. 2.3a–d. Extension of this approach to constant displacement boundary conditions, such as those done in our experiments, would in general require a numerical procedure, such as the boundary collocation method [65], to capture the interaction of the crack with the boundary. Nevertheless, as the sample size increases with respect to the crack length, the resulting crack paths approach the analytic prediction, as shown for a Gaussian bump surface in Fig. 2.6.

2.3.2   Phase-Field Model on Curved Surfaces

For modest sample sizes with constant displacement boundary loading, a numerical approach is required because of the interaction between the crack and the boundary. To predict the curved fracture trajectories, we adapt the KKL phase-field model [66, 67] to include curvature by incorporating the height profile of the substrate into the two-dimensional strain field [68]. This numerical model treats local material damage as a scalar field that evolves if there is both sufficient elastic energy density and a local gradient in the field. As depicted in Fig. 2.3c and d, these conditions are met at the tip of a propagating crack. This model captures the full crack paths, as shown by the black curves overlying experimental results in Fig. 2.3e, f.

By modulating the usual linear elastic strain energy density by a function of the phase, we begin with the strain energy density

$$\mathcal{E}_s = g(\phi) \left(\frac{1}{2} \sigma_{ij} \varepsilon_{ij} \right), \tag{2.15}$$

where $g(\phi)$ describes the softening from material to vacuum. We choose the form of $g(\phi)$ to be $g(\phi) = 4\phi^3 - 3\phi^4$, a choice consistent with [66, 69]. Note that the strain energy density recovers its usual form in intact ($\phi = 1$) regions and vanishes in broken ($\phi = 0$) regions. Given the total energy, $\int \mathcal{E} dA$, the Euler-Lagrange equations provide the governing force-balance equations for the displacements, $\mathbf{u}(x, y)$ [70]:

$$0 = \partial_\alpha \left[g(\phi)\sigma_{\alpha\beta} \right], \tag{2.16}$$

where

$$\sigma_{\alpha\beta} = \frac{E\nu}{1 - \nu^2} \varepsilon_{\gamma\gamma} \delta_{\alpha\beta} + \frac{E}{1 + \nu} \varepsilon_{\alpha\beta} \tag{2.17}$$

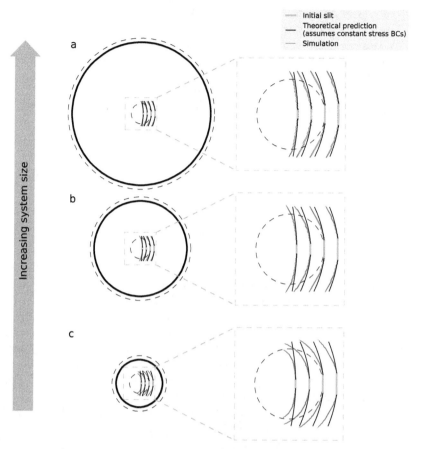

Fig. 2.6 As the size of a sheet increases with respect to the crack length, the resulting crack paths approach the analytic prediction. Contant displacement simulations performed here all have the boundary displacement chosen to induce an initial boundary stress of $\sigma_{\rho\rho}(\rho = R) = 0.02Y$ before the initial slit is introduced. The dashed black circles in the center of each sample denotes the width x_0 of the bump. Samples have diameters of 144ξ, 96ξ, and 48ξ ($14.1x_0$, $9.4x_0$, $4.7x_0$) for panels (**a**), (**b**), and (**c**), respectively, where $\xi = \sqrt{\kappa/2e_c}$ is the length scale associated with the process zone

is the local stress tensor for plane stress conditions. As for the evolution of the phase, $\phi(x, y)$, we use the form [70]

$$\mathcal{E}_\phi = \frac{\kappa}{2} (\nabla\phi)^2 + g(\phi)\left(e_\phi - e_c\right),$$ (2.18)

yielding the evolution equation

$$\chi^{-1}\frac{\partial\phi}{\partial t} = \min\left(\kappa\nabla^2\phi - g'(\phi)(e_\phi - e_c), 0\right).$$ (2.19)

The Laplacian term penalizes gradients in the phase associated with the crack faces, while the remaining terms favor successive damage in regions of high energy density. Here, e_c is a critical energy density at which it becomes favorable for ϕ to decrease. Our definition of e_ϕ in Eq. 2.19 breaks the symmetry of the standard Ginzburg-Landau form used in [66] and elsewhere in two ways. First, as in [71], we ensure that $\partial_t \phi \leq 0$, such that the damage of the material cannot be healed over time. Second, as in [69], we tune the energy density, e_ϕ, such that the material does not break as the result of compression. For our plane stress conditions, this takes the form

$$e_\phi = a \frac{E}{4(1-v)} \varepsilon_{\gamma\gamma}^2 + \frac{E}{2(1+v)} \left(\varepsilon_{\alpha\beta} - \frac{\delta_{\alpha\beta}}{2} \varepsilon_{\gamma\gamma} \right)^2 \tag{2.20}$$

where $a = 1$ in regions where $\varepsilon_{\gamma\gamma} \geq 0$ and $a \leq 0$ in regions where $\varepsilon_{\gamma\gamma} < 0$. Constant displacement boundary conditions (u_x, u_y = fixed) hold for all time. An initial $\phi(x, y)$ field is prescribed.

For a fixed curved surface defined by $z = h(x, y)$, the strain tensor of a flat material elastically confined to the surface encodes how infinitesimal distances change in the deformed body with respect to the resting state of the solid (flat, unstrained), and reads [72]

$$\varepsilon_{\alpha\beta} = \frac{1}{2} \left(\partial_\alpha u_\beta + \partial_\beta u_\alpha + \partial_\alpha h \partial_\beta h \right). \tag{2.21}$$

We then use Eq. 2.21 in Eqs. 2.16 and 2.19, approximating the metric as being flat, $g_{\alpha\beta} = \delta_{\alpha\beta}$.

We found that solving Eqs. 2.16 and 2.19 iteratively gave identical results to those of a nonlinear mixed finite element approach for sufficiently small time steps. For our iterative approach, we first write Eqs. 2.16 and 2.19 in finite difference form:

$$0 = \nabla \cdot (g(\phi_k)\sigma_{k+1}) \tag{2.22}$$

$$(\phi_{k+1} - \phi_k) = dt \, \chi \left[\kappa \nabla^2 \phi_{k+1} - g'(\phi_k) \left((e_\phi)_{k+1} - e_c \right) \right]. \tag{2.23}$$

Choosing $\nabla^2 \phi_{k+1}$ on the right hand side of Eq. 2.23 corresponds to an implicit Euler method. We write one step of this scheme in variational form and solve it using the finite element method in FEniCS [73].

While the first results from these simulations give good agreement with experiments, a systematic deviation in the extensions of the crack tips further from the bump is evident in Fig. 2.3e. In the experiments, the tip closer to the bump begins its advance first, and the dynamics of the tip are not purely quasistatic. In the phase-field simulation, simply suppressing the tip further from the bump for a short time until the near tip has reached a distance matching experiment eliminates this deviation, as shown in the inset of Fig.2.3e.

Fig. 2.7 Curvature arrests a center crack. (**a**) As the aspect ratio of the bump increases while the initial stress at the boundary $(\sigma_{\rho\rho}(R) = 0.068\,Y)$ remains fixed, the final crack length decreases. (**b**) Simulations reveal that as the aspect ratio of the bump increases, the intensity of stress concentration falls below the critical value at progressively shorter crack lengths. *Inset:* Final crack lengths from spring-lattice (squares) and phase-field simulations (triangles) mimic the arrest behavior seen in experiment (colored circles with error bars marking one standard deviation). The solid line is a guide to the eye

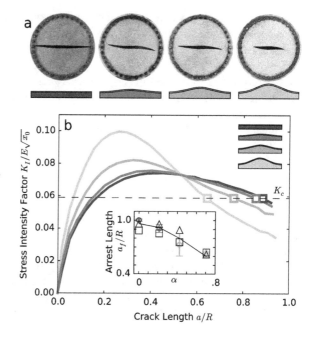

2.4   Crack Arrest

Having seen how curvature affects the initiation and propagation of cracks, we now turn our attention to the ability of curvature to arrest cracks. As seen previously in Fig. 2.3, curved cracks can terminate before reaching the sample boundary. We find, moreover, that curvature can arrest cracks even for cases in which the path is undeflected, as shown in Fig. 2.7. In flat sheets, center cracks propagate all the way to the boundary, but if we introduce a bump while holding the initial stress at the boundary fixed, the final crack length decreases.

From the decaying isotropic stress profile, we can infer that curvature generates azimuthal compression, halting the crack's advance. Using our phase-field model, we indeed find that increasing the aspect ratio of the bump lowers the intensity of stress concentration for larger crack lengths (Fig. 2.7b). A fully 3D spring network simulation using finite element methods provides additional confirmation (open squares in Fig. 2.7b). Thus, curvature decreases the final crack length, despite promoting crack initiation on top of the bump.

2.5   Controlling Cracks with More Complex Surfaces

Curvature's influence on the propagation of cracks that we have investigated on the bump is not peculiar to that surface. As shown in Fig. 2.8, we demonstrate this generality by testing a number of additional surfaces, including spherical caps

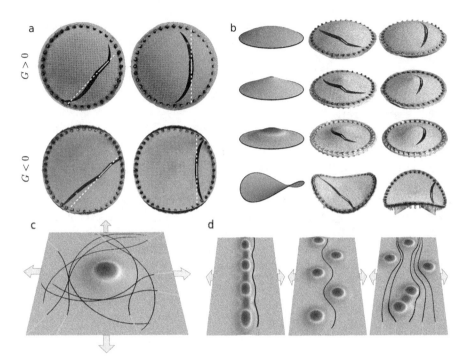

Fig. 2.8 Tuning crack paths with the curvature landscape. (**a**) Inverting the sign of the curvature (red for positive, blue for negative) inverts the behavior of the crack, as shown by the contrasting crack paths on a $L = 12$ cm spherical cap (top, $G = 1/L^2$) and on a $L = 15$ cm pseudospherical saddle (bottom, $G = -1/L^2$). Seed crack locations are marked in green. (**b**) On spherical caps, cones, and bumps, the positive integrated curvature from the center to the crack's position directs cracks towards the azimuthal direction, while the negative curvature saddle inverts this behavior. (**c–d**) Further phase-field simulations demonstrate that curvature can protect a region of a material conformed to a bump ((**c**) here under 3% biaxial displacement) or induce desired crack paths ((**d**) here shown under 1.5% uniaxial displacement). Final crack paths (black) for various initial slits (green) are overlaid to demonstrate that the bumps' central regions are protected. The results demonstrate that merely the addition of simple bumps offer a wide range of control, in experimentally realizable conformations

(uniform $G > 0$), cones ($G = G_0 \delta(\mathbf{x})$), and pseudospherical saddles (uniform $G < 0$). A region of positive curvature, such as the tip of a cone, locally stimulates crack growth near the region, but also guides cracks around that region. Conversely, negative curvature of a saddle suppresses crack growth and orients cracks away from the center (see Fig. 2.8). Thus an opposite curvature source induces an opposite response, allowing the behavior of cracks to be tuned by engineering the curvature landscape.

In Fig. 2.8c, d, we demonstrate the robustness of curvature's effects by considering samples without azimuthal symmetry using the phase-field model. Here, we use a bump to protect a central region from incoming cracks of various orientations, to produce oscillating cracks, and to focus and diverge possible crack paths. For the geometries of Fig. 2.8d, a somewhat reduced critical stress intensity factor compared to our experimental material prevents crack arrest. Though the stress is highest on top of a bump, these regions are protected from approaching cracks.

2.6   Conclusion

The use of substrate curvature to control fracture morphology differs from using existing cracks or inclusions in that our method requires no introduction of pre-existing structure into the fracturing sheets [64, 74]. For brittle sheets with isotropic elasticity, curvature-induced stresses are independent of material parameters and only dependent on geometry. Therefore, our results represent the effects of substrate curvature on fracture morphology for a wide range of materials, with potential implications for thin films, monolayers [54, 75], geological strata such as near salt diapirs [55, 76], and stretchable electronics [77]. Since the results are based on the modulations of the material's metric, they should also apply beyond conformed sheets, with metrics engineered by other methods, such as, temperature gradients [78] or differential swelling [79].

Chapter 3
Conforming Nanoparticle Sheets to Surfaces with Gaussian Curvature

Nanoparticle monolayer sheets are ultrathin inorganic-organic hybrid materials that combine highly controllable optical and electrical properties with mechanical flexibility and remarkable strength. Like other thin sheets, their low bending rigidity allows them to easily roll into or conform to cylindrical geometries. Nanoparticle monolayers not only can bend, but also cope with strain through local particle rearrangement and plastic deformation. This means that, unlike thin sheets such as paper or graphene, nanoparticle sheets can much more easily conform to surfaces with complex topography characterized by non-zero Gaussian curvature, like spherical caps or saddles. Here, we investigate the limits of nanoparticle monolayers' ability to conform to substrates with Gaussian curvature by stamping nanoparticle sheets onto lattices of larger polystyrene spheres. Tuning the local Gaussian curvature by increasing the size of the substrate spheres, we find that the stamped sheet morphology evolves through three characteristic stages: from full substrate coverage, where the sheet extends over the interstices in the lattice, to coverage in the form of caps that conform tightly to the top portion of each sphere and fracture at larger polar angles, to caps that exhibit radial folds. Through analysis of the nanoparticle positions, obtained from scanning electron micrographs, we extract the local strain tensor and track the onset of strain-induced dislocations in the particle arrangement. By considering the interplay of energies for elastic and plastic deformations and adhesion, we construct arguments that capture the observed changes in sheet morphology as Gaussian curvature is tuned over two orders of magnitude. This chapter is adapted from [44] with permission from The Royal Society of Chemistry.

© Springer Nature Switzerland AG 2020
N. Mitchell, *Geometric Control of Fracture and Topological Metamaterials*,
Springer Theses, https://doi.org/10.1007/978-3-030-36361-1_3

3.1   Gaussian Curvature and Nanoparticle Sheets

While any flat thin sheet can easily be rolled into a cylinder, common experience suggests that conforming the same sheet to a sphere is considerably more difficult. In order to accommodate the curvature of the sphere, one must fold, cut, or stretch the sheet. On surfaces with Gaussian curvature—that is, curvature in two independent directions, such as on a sphere or saddle—triangles no longer have interior angles which sum to 180°. Conforming a flat sheet tightly to such a surface thus necessarily introduces stresses from stretching or compression. If the stresses build up, the material may respond by delaminating or forming cracks, dislocations, or folds [14, 43, 50]. For applications where initially flat sheets are to conform to arbitrary surface topographies, the ability to cope with Gaussian curvature therefore translates into the ability to bend and deform locally in-plane.

Relatively stiff materials such as paper or graphene have difficulty coping with these stresses, and therefore rip or fold instead of conforming to surfaces with Gaussian curvature. Studies of softer elastic sheets, on the other hand, have led to the understanding of curvature as a tool for patterning defects [9, 10, 14], cracks [43], folds [80, 81], wrinkles [81, 82], blisters [50], and even controlling phase transitions to and from the solid state [12, 13]. In this article, we extend these efforts by focusing on a particular material: close-packed nanoparticle monolayers. These hybrid organic-inorganic materials combine remarkably high Young's modulus (several GPa) with the ability to deform and rearrange locally in a plastic manner. Furthermore, their versatility has given rise to prospective applications in filters [83], solar cells [84], sensors [85–87], batteries [88], and beyond due to their optical [89], electrical [90, 91], and chemical properties [92].

In nanoparticle monolayers, individual metallic or semiconducting particle cores are embedded in a matrix of interpenetrating ligand molecules that are bound to each core [93, 94], with the organic matrix largely determining the sheet's bulk mechanical properties. While these properties have been studied for sheets in planar geometries [95–97] and for cylindrical, scroll-like structures [98], the ability of flat sheets to conform to surfaces with Gaussian curvature has received little attention [54]. Here, we investigate this by stamping monolayers of dodecanethiol-ligated gold nanoparticles onto surfaces formed by lattices of larger polystyrene (PS) spheres.

The situation we address begins with pre-assembled flat sheets that deform as they are stamped against a highly curved surface, as illustrated in Fig. 3.1. For nanometer-thin sheets, van der Waals forces generate adhesion that effectively immobilizes the nanoparticles as they come into contact with the substrate. Furthermore, in contrast to continuum elastic sheets, the discrete nanoparticle lattice allows for the formation and proliferation of defects in addition to straining, folding, and fracturing during the conformation process.

The effect of strong pinning to the substrate results in strikingly different behavior from that of equilibrium arrangements of interacting Brownian particles on spheres [10, 13], frustrated equilibrium conformations of macroscopic, continuum

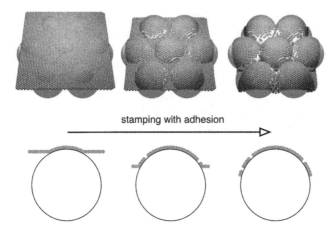

Fig. 3.1 Nanoparticle sheets conform to highly curved surfaces. In the situation under study, a preformed nanoparticle monolayer is pressed against a substrate comprised of a lattice of larger spheres. As the sheet is stamped, the nanoparticles become pinned to the substrate spheres. The three snapshots (top) are from a simulation of an elastic network. As the thin sheet conforms to the substrate while experiencing pinning forces, stresses result in broken bonds between nanoparticles

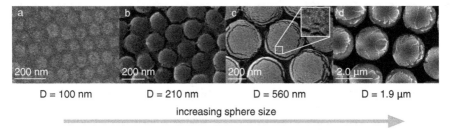

Fig. 3.2 Sphere size controls the morphology of stamped nanoparticle sheets. (**a**) At small sphere diameter D, the monolayer sheet is able to cover the polystyrene sphere array completely, but does not fully conform to each sphere. (**b–c**) As D increases, the sheets tightly conform to the upper portions of the spheres. However, they no longer bridge the crevices between spheres and instead form azimuthal cracks. (**d**) At even larger D, sheets buckle out of plane, creating radial folds

elastic sheets [43, 50], or non-equilibrium growth of colloidal crystals on spherical interfaces [12]. Because the pinned sheet cannot relax to minimize free energy, the effects of geometric frustration build up according to history-dependent, sequential rules. This sequential adhesion gives rise to qualitatively different stress fields in the sheet and suppresses wrinkling before the appearance of sharp folds.

Depending on the Gaussian curvature, G, of the corrugated substrate, which we control by the PS sphere diameter D via $G = 4/D^2$, we find three characteristic stamped-sheet morphologies. As seen in Fig. 3.2, increasing D leads from sheets that entirely cover the corrugated substrate to sheets that have fractured into caps closely conforming to the top portions of the PS spheres. Finally, the largest PS spheres yield caps exhibiting radial folds similar to those seen in macroscopic, continuum sheets [82]. We show that these curvature-dependent morphologies emerge from the interplay between strong pinning to the substrate, elastic energies, and costs for defect formation. This allows us to generate predictions for the conditions required to obtain full coverage and for the limits to which nanoparticle sheets can conform tightly to arbitrarily curved surfaces.

In what follows, we first describe the experiments and resulting sheet morphologies. We then provide energy scaling arguments that rationalize the crossovers between stamped sheet morphologies as a function of D or G. In subsequent sections, we examine each regime in turn and find that detailed measurements corroborate the overall scaling picture. We directly measure the local strain within the stamped sheets and compare them to simulations of two-dimensional spring networks made to conform to sphere lattices. From these measurements and simulations, we determine the onset of finite size effects due to the discrete nature of the nanoparticles. This analysis provides a correction to the overall scaling picture for small PS sphere sizes and allows us to predict the maximum polar angle up to which the sheet can tightly conform to individual PS spheres without material failure.

3.2   Experimental Procedure

Dodecanethiol-ligated gold nanoparticles were synthesized via a digestive ripening method followed by extensive washing with ethanol and finally dissolution in toluene [99]. This process yielded nanoparticles with diameter 5.2 ± 0.3 nm and ligand lengths 1.7 ± 0.3 nm. Nanoparticle monolayers were self-assembled at the surface of a water droplet. After depositing a drop ($\sim$150 μL) of deionized water onto the hydrophobic surface of a piece of polytetrafluoroethylene (PTFE), 5–7 μL of the nanoparticle-toluene solution were pipetted around the drop perimeter. The solution climbed to the top of the droplet almost immediately, and, as the toluene evaporated, the nanoparticles self-assembled into a close-packed monolayer with a lattice spacing of 7.2 ± 0.8 nm (Fig. 3.3a–d). Waiting several hours allowed some of the water to evaporate as well. Given the strong pinning of the drop's contact line to the substrate, this evaporation changed the droplet shape from a spherical cap to a flattened dome (not shown in Fig. 3.3b).

At this stage, a silicon chip coated with a lattice of polystyrene (PS) spheres was gently pressed against the assembled monolayer and peeled away (Fig. 3.3e, f). These PS sphere lattices were created by diluting solutions of PS spheres (Bangs Laboratories) by a factor of 100 using deionized water, then depositing 5–7 μL of the diluted solution onto 25 mm^2 silicon chips and allowing them to dry. Our experiments used sphere diameters ranging from 100 nm to 1.9 μm. Variations in PS sphere sizes increased with their diameters, ranging from a standard deviation of 2% for 100 nm spheres to 12% for 800 nm spheres, while the 1.9 μm spheres had a standard deviation of 20%. Because the nanoparticle monolayers readily adhere to

Fig. 3.3 Schematic of the experimental procedure for conforming self-assembled gold nanoparticle monolayer sheets to a lattice of polystyrene spheres. (**a–b**) Drying-mediated assembly of a nanoparticle monolayer at the surface of a water droplet. (**c–d**) Close-up illustrating the self-assembly of the monolayer at the water-air interface. (**e–f**) Stamping a lattice of larger polystyrene (PS) spheres onto the nanoparticle monolayer and peeling it away from the water droplet

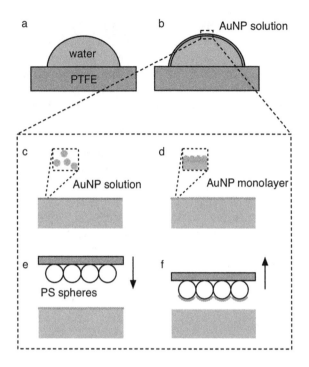

the PS spheres, the layers delaminate from the water and transfer to the PS spheres, as when inking a stamp. These 'stamped' monolayers were then imaged using a Carl Zeiss Merlin scanning electron microscope (SEM).

3.3 Monolayer Morphology: Coverage, Cracks, and Folds

SEM imaging revealed that the nanoparticle sheets reproducibly retain their monolayer structure as they are transferred onto the substrate of PS spheres. The sheet morphology, however, varies with the size of the PS spheres used. For PS diameters $D \approx 100\,\text{nm}$, monolayers typically cover the substrate without cracks or folds (Fig. 3.2a). For these small D, the monolayers do not enter deeply into the crevices between spheres, instead getting pinned at the apex of each PS sphere and bridging the crevices as freestanding membranes.

Once D becomes larger, the stamped sheets are able to follow the substrate surface topography more closely, creating snugly fitting caps. Remarkably, the sheets conform tightly to the PS spheres up to polar angles of 20–30° (measured from the apex of each sphere) without buckling, wrinkling, or creating folds. This already indicates behavior quite distinct from that of other thin sheets, such as paper, mylar, polystyrene, or graphene, which invariably generate folds or rip [50, 80, 100–102].

At larger polar angles, azimuthally oriented cracks appear, which hint at large radial stress as the sheets conform to the PS spheres during the stamping process. These cracks prevent the sheets from bridging the gap between neighboring spheres (Fig. 3.2b, c). For sphere diameters larger than roughly 1 μm, not only do the sheets tear azimuthally to form caps on each sphere, but also they form localized radial folds to accommodate the mismatch between flat and spherical metrics (Fig. 3.2d).

The azimuthal cracks in Fig. 3.2b, c and the radial folding lines in Fig. 3.2d form during the stamping process, in which the monolayers are deformed under vertical pressure to conform against the non-Gaussian topography, as sketched in Fig. 3.1. Once the nanoparticles are in contact with the polystyrene surface, the adhesion immobilizes these local deformations. For D around 200 nm, portions of the monolayer that did not adhere to PS spheres tend to tear in the interstices between polystyrene spheres. For larger D, the azimuthal fractures become more pronounced, allowing the interstitial portions of the sheet to recede further down (Fig. 3.2c). For the largest sphere sizes ($D \geq 690$ nm), the non-adhering portions may be swept away as the water dewets the chip while it is being pulled off the droplet at the end of the stamping process (Fig. 3.2d).

3.4 Energy Scaling

In this section, we provide a self-consistent rationalization for the observed changes from incomplete adhesion to plastic deformation to folding, using scaling arguments for continuum sheets. In subsequent sections, we examine each regime in turn and find that detailed measurements corroborate the overall scaling picture presented here, while also providing corrections due to the discrete lattice structure of our sheets.

A simple geometric insight underpins the trend in behavior seen in Fig. 3.2. On a flat sheet, the circumference of a circle grows in proportion to its radius, r. On a sphere, however, the circumference of a circle at the same distance r from the sphere's apex grows more slowly due to the Gaussian curvature. In other words, when a flat disc of given r is made to conform to the surface of a sphere, it must deform to compensate for the deficit in circumference. The sheet must therefore not only bend, but also strain elastically in the form of radial expansion, azimuthal compression, or some combination of the two.

If the sheets furthermore become pinned to the PS spheres during the stamping process, the nanoparticles attach sequentially one annulus at a time, starting from each sphere's apex (Fig. 3.4a). As successive annuli conform to the substrate, the cost of elastic energy may exceed the energetic costs associated with delaminating, forming defects, ripping apart, or folding. To understand the competing energy scales, consider an annulus of nanoparticle sheet with radial width δr that has been conformed onto a PS sphere of diameter D to sit at polar angle θ. Such an annulus has an area $\pi D \delta r \sin\theta$ (to zeroth order in strain). Conforming this

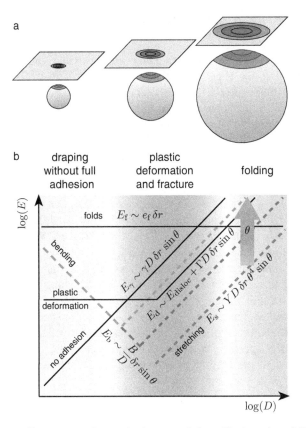

Fig. 3.4 Energy scaling captures changes in sheet morphology. The interplay of different energy costs provides crossovers from fully covered PS lattices (incomplete adhesion, green region), to plastic deformation (red region), to the formation of localized folds (blue region). Each energy is for a nanoparticle annulus of radial width δr—with stiffness Y and bending modulus B—and a PS sphere of diameter D. The energy cost of *not* adhering to the PS substrate, E_γ, grows with the area of the annulus, $\pi D \sin\theta \delta r$, and depends on the adhesion energy, γ. Similarly, the stretching energy, E_s, and the energy of plastically deforming the annulus by dislocation proliferation, E_d, likewise grow with the area of the annulus. The stretching energy also depends strongly on the polar angle, θ, through the strain $\epsilon_{ij} = \epsilon_{ij}(\theta)$ as $E_s \sim Y D \delta r \theta^4 \sin\theta$, depicted by the offset between colored dashed blue, gray, and orange lines. The plastic deformation energy, E_d, has a minimum set by the energy of unbinding a pair of dislocations, E_{disloc}, and the factor Γ is a phenomenological constant characterizing the work necessary to plastically deform a unit area of the sheet. The energy of creating a localized fold, E_f, is set by the energy to crease the sheet. The fold energy per unit length of the fold, e_f, depends on the fold angle and microscopic details of the lattice

annulus to the sphere requires energies due to bending and stretching, and these conformational energy costs compete with alternative behaviors, such as remaining free-standing instead of conforming, plastically deforming and fracturing, or folding.

3.4.1 Energy Costs to Conform: Bending and Stretching

First, conformation of the annulus requires areal bending energy density $\mathcal{E}_b \sim B/D^2$, where B is the sheet's 2D bending modulus. The total bending energy in the annulus then becomes $E_b \sim (B/D)\delta r \sin\theta$. Here we are neglecting small corrections to this approximation of order $\mathcal{O}(\theta^2)$. Thus, the cost of bending decreases as D grows, as shown by the downward dashed line in the left portion of Fig. 3.4b.

Second, the sheet must also stretch to conform to a sphere. The total stretching energy, E_s, stored in the annulus is proportional to its surface area and the stretching energy density. This stretching energy density, $\mathcal{E}_s$, is a quartic function of polar angle on the sphere, $\mathcal{E}_s \sim Y\theta^4$, as shown in Appendix B. Here, Y is the stiffness of the sheet. Therefore, the cost of stretching increases linearly with D, but the magnitude depends sensitively on the polar angle: $E_s \sim YD\delta r\,\theta^4 \sin\theta$. While Fig. 3.4b omits linear and sublinear dependence on θ for clarity, this strong dependence of E_s on polar angle is shown by the rising dashed lines. The changing colors (blue, gray, orange) denotes that, for a given sphere size D, the stretching energy in an annulus grows rapidly with polar angle.

We emphasize that the stretching energy scaling in our sheets strongly contrasts from the well-studied case of equilibrated sheets conformed to a sphere, in which the energy density *decreases* quadratically with polar angle, θ, for small θ. This difference highlights the distinct character of sequential adhesion to a substrate seen in our system.

3.4.2 Alternatives to Elastic Conformation: Avoiding Adhesion, Plastic Deformation, and Folding

These elastic energies compete with the possibility of adopting alternative behaviors. Instead of elastically bending and stretching to conform, the sheet may only partially conform to the sphere, or it may plastically deform, rip apart, or form folds.

While stretching and bending cost energy, the adhesion process can *relieve* energy as well, since it replaces two interfaces (nanoparticle-air and air-PS) with a single one (nanoparticle-PS). This replacement relieves energy in proportion to the area of adhered material, so there is a fixed areal energy density $\mathcal{E}_\gamma$ relieved by adhering to the PS sphere. For the annulus, this translates into a total cost of *not* adhering to the substrate, $E_\gamma \sim \gamma D\delta r \sin\theta$, that increases linearly with D. Here, γ is the areal surface energy density relieved by adhering a nanoparticle sheet to the PS substrate.

While the stretching energy scales as $E_s \sim D \sin\theta\delta r\,\theta^4$, the energy cost E_d of relieving stress through plastic deformation of the annulus scales similarly with sphere diameter, but has a far weaker scaling in θ: $E_d \sim \max(E_{\text{disloc}}, \Gamma D\delta r \sin\theta)$, where E_{disloc} is the energy of unbinding a single pair of dislocations and Γ is a phenomenological factor capturing the work required to damage a unit area of the material. The minimum possible energy to create the first defect pair, E_{disloc}, sets the lower cutoff that freezes out defect proliferation at small D. E_{disloc} is determined by the core energy of a dislocation and the elastic cost of deforming the portion of sheet surrounding the dislocations, which depends on microscopic features of the lattice. Finally, the energy cost for creating a fold in the sheet, E_f, increases only with the fold length ($E_f \sim e_f \delta r$, where e_f is the fold energy per unit length) and thus is independent of D.

3.4.3 Three Regimes Arise from Energy Scaling

Figure 3.4b represents these energy scaling relations schematically. Throughout this figure, linear and sub-linear dependences on the polar angle θ are suppressed for clarity. In particular, the adhesion and bending energies grow as $\sin\theta$, and we omit this dependence. Conversely, we do include the strong θ dependence of the stretching energy, and illustrate this strong dependence by the colored dashed lines.

From this scaling we infer that for sufficiently small sphere sizes (or, equivalently, large Gaussian curvature), the lowest cost will be incurred by incomplete adhesion, as this causes the least distortion in the flat sheet. The green region in Fig. 3.4b represents this regime, which corresponds to the experimental results in Fig. 3.2a.

For larger sphere sizes, bending becomes energetically cheaper than not adhering. However, in order to conform tightly to the sphere, the monolayer needs to not only bend, but also stretch or compress. For annuli at small polar angles θ, this elastic energy cost can be negligible, but as θ grows for a given D, the cost will eventually exceed the penalty for creating defects. As a result, beyond some critical polar angle θ_c, plastic deformation in the sheet will cause a proliferation of dislocations. We expect that the formation of cracks follows as a result of this defect formation, along with the tension that remains while defects are formed. Since the in-plane stretching is tensile along the radial direction, as we will see, cracks open up along the azimuth, perpendicular to the radial tension. This regime is represented by the red region in Fig. 3.4b and corresponds to the experimental results in Fig. 3.2b, c.

For the largest PS sphere sizes, yet another crossover occurs due to the difference in scaling between the costs for either elastic stretching or plastic deformation, which increase linearly in D, and the costs of forming localized folds, which is independent of D. This is the regime shown in blue in Fig. 3.4b, corresponding to Fig. 3.2d. Because the energy cost for fold formation lies below that of plastic deformation in the blue regime, the first response as strains build up will be to form folds rather than the proliferation of dislocations.

This energy scaling captures all three regimes of stamped nanoparticle sheet morphology seen in Fig. 3.2. We note that this framework operates in the continuum limit. Additionally, our picture assumes that chemical properties of the polystyrene do not vary with PS sphere size, an effect that could alter the adhesion energy in Fig. 3.4b. Nevertheless, the essential features are supported by quantitative comparisons with experiments and simulations given in the following sections.

In the remaining sections, we discuss in more detail each of the mechanical responses of the flat sheets to the enforced geometric mismatch: bending, stretching, dislocation proliferation, crack formation, and folding.

3.5   Bending and Adhesion

The crossover from incomplete adhesion to full adhesion with plastic deformation occurs in our experiments for PS spheres with diameters $D \approx 200\,\text{nm}$. This crossover enables an estimate of the bending rigidity in nanoparticle membranes.

The two-dimensional bending energy density of a thin plate in plane stress is [72]

$$\mathcal{E}_B = \frac{B}{2}\left[\left(\nabla^2 h\right)^2 + 2(1-v)\left\{ \left(\frac{\partial^2 h}{\partial x \partial y}\right)^2 - \frac{\partial^2 h}{\partial x^2}\frac{\partial^2 h}{\partial y^2} \right\} \right], \tag{3.1}$$

where $h(x, y)$ is the out-of-plane displacement of the plate and B is the bending modulus. Since we are interested in the behavior near the apex of a sphere, we Taylor expand around $\theta = 0$ to obtain

$$\mathcal{E}_B = \frac{B}{R^2}\left[(v+1) + 2(v+1)\theta^2 + \mathcal{O}\left(\theta^4\right) \right] \tag{3.2}$$

$$\approx \frac{4B(v+1)}{D^2}, \tag{3.3}$$

where in the second line we dropped corrections to the bending energy that scale quadratically with the polar angle, θ. We take the Poisson ratio to be $v = 1/3$, the value for a triangular lattice of spring-coupled nodes, in accordance with the measured value for nanoparticle sheets [103]. We take an average radius of curvature of $D/2 \approx 100$ nm for the crossover.

At the small-sphere crossover between incomplete adhesion and plastic behavior, we should expect the bending energy to match the adhesion of the nanoparticle sheet with polystyrene. Using the result of Ref. [104], we estimate the adhesion energy from the surface tensions of dodecane (21 mN/m) and water (72 mN/m), the surface energy of solid polystyrene ($\sim$42 mN/m) [105], and the molar volumes of each. The result is an adhesion energy of $\gamma_{PS} + \gamma_{\text{dodecane}} - \gamma_{PS,\text{dodecane}} \approx 60$ mN/m. We expect that the bending energy, $\mathcal{E}_B$, matches this value at the crossover. This gives a bending modulus for the nanoparticle sheets of $B \approx 4.5 \times 10^{-16}$ Nm.

From this we may deduce a lower bound on the effective thickness t_{eff} of the sheet, which can deviate from the physical thickness due to the non-continuum nature of the material [98]. The bending modulus is related to t_{eff} via $B = Yt_{\text{eff}}^2/12(1-v^2)$. Here the 2D stiffness $Y = Et$ is the product of Young's modulus

E and physical thickness t. If we assume $E \sim 3$ GPa, as is appropriate for fully dried monolayers [95, 106], we obtain $t_{\text{eff}} \approx 14$ nm, about 60% larger than the physical thickness of $t \approx (d_{\text{Au NP}} + 2 \times \ell_{\text{ligand}})$ nm $= 8.2$ nm. However, we expect that during the stamping process there is residual water embedded in the ligand matrix. The presence of water molecules in the matrix has been shown to drastically affect the elastic properties, reducing elastic moduli by potentially several orders of magnitude [107, 108]. Such decrease in E then implies an increase in t_{eff}, possibly up to around $10t$ as observed for dried monolayers [98].

The crossover from incomplete adhesion on small spheres to tightly conforming to larger spheres is reminiscent of the crossover in a thin sheet's 'bendability', which is the ratio of tensile to bending forces, TW^2/B, where T is the tension at the edge of a sheet of width W due to in-plane stretching or interfacial forces [82]. If we consider the case where $W \sim D$, so that the sheet covers the same proportion of the sphere for different sphere sizes, then as the PS sphere size increases, so too does the bendability of the sheet. Our system differs from these recent studies of comparably stiff sheets, however, because of the strong pinning of the nanoparticle sheet to the substrate. The apparent force imbalance in the stretching of the sheet measured in simulations shows that adhesion enables a disproportionate increase in radial tension, at a rate faster than long-range elasticity would allow. Specifically, adhesion supplies a tension which offsets the imbalance of in-plane stresses, $\partial_r(r\sigma_{rr}) - \sigma_{\phi\phi}$. While this quantity would vanish without pinning, here the stress imbalance grows as θ^2 for small to moderate polar angles.

3.6 Strain Analysis

During the stamping process, the first contact between the nanoparticle sheet and a PS sphere occurs at the sphere's apex, $\theta = 0$, where the sheet will be pinned. Subsequent annuli of the sheet will need to strain or undergo plastic deformation in order to conform tightly to the surface of the PS sphere, but once this has occurred, these annuli also will become pinned to the polystyrene. This means that we can obtain information about the local strain by using the individual nanoparticles as markers and extracting differences in their average spacing along a sphere's surface. Given the random disorder inherent already in the flat sheets, this procedure requires ensemble averages over several different imaged PS spheres for statistically relevant results.

3.6.1 Image Analysis

To study the strains and defect densities of nanoparticle sheets, we use a custom image analysis routine on each SEM image to identify the nanoparticle locations and to identify the nearest-neighbor connectivity of the nanoparticle lattice [109].

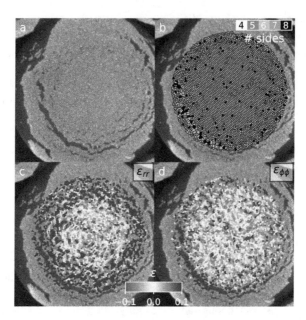

Fig. 3.5 Identification of defects and extraction of the local strain tensor. (**a**) Nanoparticles are identified in the original SEM image. (**b**) Using a Voronoi tessellation, we enumerate the neighbors of each nanoparticle. For each nanoparticle with six neighbors, comparing the Voronoi cell to a regular hexagon lying on the tangent plane of the sphere yields the strain tensor. To restrict the analysis to elastic deformations, we omit particles whose Voronoi cell is deformed well beyond the elastic limit of the material, keeping only hexagons whose perimeter to surface area ratio, $s \equiv P/\sqrt{A}$, satisfies $s < s_{\text{cutoff}} = 3.8$. (**c–d**) The radial strain in the sheet, ϵ_{rr}, increases with distance from the apex, while azimuthal strain, $\epsilon_{\phi\phi}$, does not

We bandpass each image in two steps: first convolving it with a Gaussian (whose parameters include nanoparticle characteristics such as lattice spacing) and then convolving the result with a boxcar function. Subtracting the two gives a high-pass-filtered image from which we extract particle positions.

A Delaunay triangulation provides the lattice topology and the nearest neighbors for each particle. Defects in the lattice are particles with fewer than six or greater than six neighbors (disclinations), and pairs of oppositely signed disclinations form dislocations (for example, a 5–7 disclination pair). Figure 3.5b shows an example Voronoi tesselation of a triangulated nanoparticle sheet draped on a 690 nm diameter PS sphere. The Delaunay triangulation also enables a direct measurement of the local strain tensor, ϵ_{ij}. For particles with exactly six neighbors, we measure the displacements of its neighbors from a regular hexagon with bonds of unit length. In this step, we account for the non-planar geometry of the substrate by computing displacements only in the tangent plane to the underlying PS sphere. By comparing each triad of the central particle and two adjacent neighbors to an undeformed reference triangle, we obtain a strain tensor for that triad of nanoparticles. For each particle that is not a defect, the average strain field of its six shared triangles

Fig. 3.6 Strain analysis shows qualitative agreement between experiments and simulations. Data from nanoparticle sheets on 62 imaged PS spheres of different diameters reveals that the radial strain, ϵ_{rr} increases with polar angle, while the azimuthal strain, $\epsilon_{\phi\phi}$, is compressive and comparatively small. The incompressible solution does not fit as well to the data, showing that nanoparticle sheets behave elastically

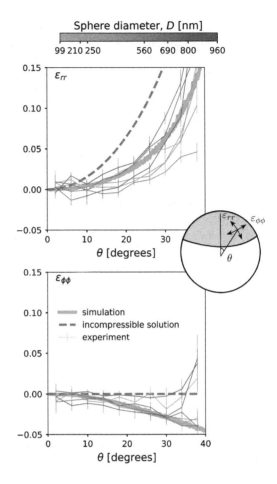

represents a measure of local strain. This strain measurement is well-defined only for particles that have six nearest neighbors—that is, those particles which do not form topological defects in the lattice.

Identifying the center of the PS substrate spheres by fitting their profile to a circle, we rotate the strain field ϵ_{ij} into polar coordinates (ϵ_{rr}, $\epsilon_{r\phi}$, $\epsilon_{\phi\phi}$) and average annular bins (i.e., bins of $\phi_i < \phi < \phi_{i+1}$) to obtain curves for $\epsilon_{rr}(\theta)$ and $\epsilon_{\phi\phi}(\theta)$ as a function of polar angle on a sphere. Typical results are shown in Fig. 3.5c, d. Figure 3.6 shows strain curves averaged over several spheres and images for each sphere size. To further reduce noise from voids and defects, we also omit particles whose Voronoi cells are deformed well beyond the elastic limit of the material. Specifically, we enforce a cutoff in the shape parameter s, defined as the ratio of the perimeter of the hexagon to the square root of its surface area, $s \equiv P/\sqrt{A}$. Here, we use the cutoff $s < s_{\text{cutoff}} = 3.8$, which removes outliers subject to more than 17% pure shear.

Figure 3.6 shows the average strain tensor components as a function of polar angle for different sphere sizes. The analysis indicates that the sheet's radial tension grows substantially, while the strain along the azimuth of the PS sphere is weakly compressive. The shear strain averages to zero, as predicted by the symmetry of the spherical geometry, with variations in the measured mean shear of <1%. As mentioned above, the nanoparticle sheets' inherent disorder creates a distribution of strain component values for each binned annulus. These distributions have a standard deviation of ~10% strain—significantly larger than the strains themselves for all but the largest values of θ considered. By averaging the strains in annular bins on each PS sphere and by performing ensemble averages over different spheres, the disorder on the scale of individual nanoparticles is largely averaged out. As Fig. 3.6 shows, these ensemble-averaged data can show quantitative differences as the PS sphere diameter D is varied. This likely is due to slight, unavoidable variations in the sample preparation conditions. However, within this variability we find no clearly discernible trends as a function of D. Considered in aggregate, these data can therefore be used for qualitative comparison with models, as we discuss next.

3.6.2 Spring Network Simulations

To gain insight into the elastic behavior during the stamping process, we model the nanoparticle sheet as a flat, triangular spring network. Simulations of such networks pinned to a lattice of spheres reproduce the trends in strain observed in the experiments (Fig. 3.6).

The simulations proceed by minimizing the free energy of a triangular spring network at each time step using a conjugate gradient method as we deposit the network onto a lattice of spheres. Whenever a node of the spring network makes contact with a substrate sphere, we irreversibly pin that node to the point of contact for the remainder of the simulation. Increasing the radii of the substrate spheres with respect to the bond length by a factor of two (and, proportionately, scaling the number of nanoparticles by a factor of four) gave virtually identical results for the strain plots given in Figs. 3.6 and 3.7, indicating that the simulations are representative of the continuum limit. Study of the finite size scaling shows that the strain curves deviate significantly from the continuum limit only for substrate sphere sizes below $D \lesssim 10a$, where a is the lattice spacing.

In the simulations, a sheet began at a distance $R = D/2$ above the plane containing the centers of the substrate spheres, each of diameter D. The network was then lowered in small increments $(0.001D)$ and the free energy was minimized for that configuration, subject to the constraint that all particles (nodes of the spring network) must lie in the common membrane plane or on a sphere, whichever is higher in the z dimension. For each step, a sequence of random kicks were applied to each node to escape local minima in the energy landscape. At the end of the relaxation process, nodes in contact with a substrate sphere—that is, within a small

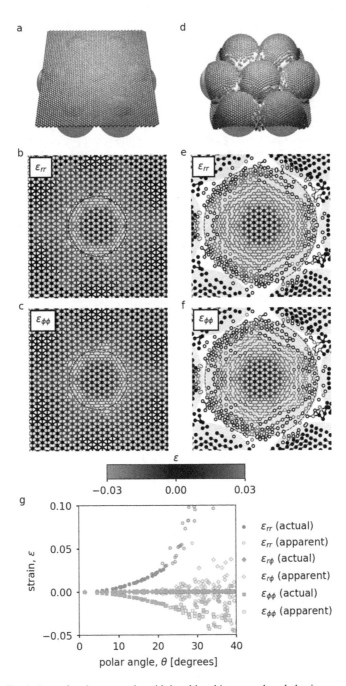

Fig. 3.7 Simulations of spring networks with bond breaking reproduce behavior seen in experiment. Spring networks were made to conform to a lattice of spheres, as in Fig. 3.1. Bonds with $\geq 3\%$ strain are removed at each time step, mimicking bond breakage. (**a–c**) As a flat, triangular spring network is pressed against an array of spheres, each node is immobilized upon contact with a substrate sphere. As the network conforms, strains build up, leading to bond breaking for polar angles larger than $\theta \sim 23°$. Particles with severed bonds are colored white at their centers in the strain images. (**d–f**) Layers of bonds continue to adhere to the substrate with many radial bonds broken. (**g**) Though the actual strains in the network's springs do not exceed 3%, the apparent strain inferred from the placement of nanoparticles continues to increase in the damaged annuli

threshold of $10^{-5}a$, where a is the rest bond length (lattice spacing)—are marked as immobilized for the remainder of the simulation.

As shown by the blue curves in Fig. 3.6, these simulations of perfectly elastic triangular networks show similar behavior in both ϵ_{rr} and $\epsilon_{\phi\phi}$ as a function of polar angle on the underlying sphere. As the membrane begins to conform to the sphere lattice, pinning ensures that the apex of the sphere experiences negligible strain, as expected. The radial stress increases quadratically, while a compressive azimuthal stress builds up more slowly. The deviation of $\epsilon_{\phi\phi}$ between experiment and simulation at large θ is due in part to the material failure and plastic deformation of the actual sheets, which is suppressed in the simulations we show in Fig. 3.6.

We note that in experiment, the nanoparticle membrane may not be perfectly flat in the interstices of the PS spheres, as the pressure of the water during stamping may push the sheet into the interstices. Modifying the simulation geometry to enforce an indentation of the sheet into the interstices of the PS lattice has only a weak effect leading to somewhat elevated strains in the final, pinned state without changing the qualitative strain behavior.

3.6.3 Comparison with Incompressible Solution

Considering the limit in which the nanoparticle sheet is incompressible allows for a useful point of reference against which we can compare the iterative adhesion of nanoparticle annuli. The strains required to conform to the substrate in this limit are indicated by the green dashed line in Fig. 3.6. Namely,

$$\epsilon_{rr} = \sqrt{\frac{R^2}{(R^2 - r^2)}} - 1, \tag{3.4}$$

where $R = D/2$ is the radius of the PS sphere, while $\epsilon_{\phi\phi} = 0$ due to incompressibility. All data, whether experimental or simulation-based, lie below this solution for ϵ_{rr}. This clearly indicates compressible behavior of our nanoparticle sheets.

3.6.4 Azimuthal Cracks in Simulations

The material cannot stretch elastically without bound: sufficiently large strains will plastically deform the sheet, severing bonds between nanoparticles to form cracks or dislocations. Indeed, the radial strains seen in Fig. 3.6 greatly exceed the critical strain for failure in flat nanoparticle membranes [99]. While we will consider plastic deformation in the next section, we note that introducing failure into the spring network simulations generates qualitatively similar morphologies to those seen in experiment. Figure 3.7 demonstrates that introducing a nominal breaking strain of 3% leads to the formation of partially intact annuli separated by azimuthal cracks. In Fig. 3.7g, we show both the strains of particles with all original bonds intact (closed markers) as well as the 'apparent' strain (open markers) resulting from triangulating the point pattern and including all particles with six nearest neighbors, regardless of whether the bonds connecting them have severed. This gives strains that remain qualitatively similar to those seen in experiment, with increased scatter in the apparent strains frozen into the broken regions pinned to the substrate.

3.7 Plastic Deformation

Given that a flat nanoparticle lattice forms a close-packed array of hexagons, any particles that do not have six nearest neighbors are defects. We record the location of each defective particle and its number of nearest neighbors. Figure 3.5b shows the Voronoi tessellation of one representative lattice overlaying the original SEM image. Each yellow site corresponds to a nanoparticle having six nearest neighbors (i.e., a hexagon), while defects are colored white, blue, green, and black for coordination numbers of $z = 4, 5, 7$, and 8, respectively.

As the sheet begins to respond with plastic deformation, dislocations proliferate in the material. The density of dislocations correspondingly increases with polar angle on a sphere, as can be seen in Fig. 3.5b. We observe that azimuthal cracks form only beyond the point of dislocation proliferation, which suggests that the material yields plastically before cracks coalesce.

3.7.1 Formation of Dislocations

The scaling arguments presented in Fig. 3.4, which operate in the continuum limit, predict that plastic deformation should be favorable at a critical angle independent of sphere diameter D. In our experiments, however, we observe an increase in the polar angle at which dislocations appear for the smallest PS sphere sizes, shown in Fig. 3.8. This observation implies that the discrete structure of the nanoparticle monolayers can be important in determining the details of their mechanical behavior. The continuum limit description of Fig. 3.4 does not include microscopic details, and therefore predicts a size-independent critical angle for the onset of plasticity. If the discrete structure of the sheet comes into play, we expect a correction to this picture to appear at small sphere sizes, where the lattice spacing is a non-negligible fraction of the system size.

As expected, the most prominent types of strain-induced defects in the nanoparticle arrangement are dislocations—i.e., pairs of Voronoi cells with 5 and 7 sides. Figure 3.8a shows a representative measurement of the crossover from low to high defect density as a function of polar angle, θ. These data were obtained

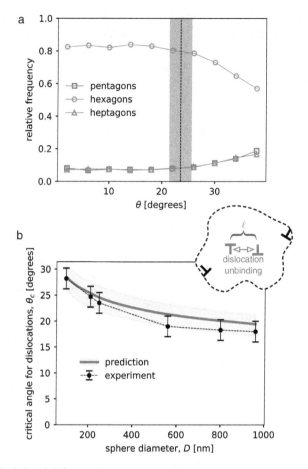

Fig. 3.8 Strain-induced defects in the nanoparticle sheets reveal non-continuum behavior. (**a**) The proliferation of defects results in increasing dislocation frequency (and correspondingly, to a decreasing frequency of hexagons) as a function of polar angle, θ. An example of the angle-dependence of defect densities is shown for nanoparticle sheets conformed to 250 nm PS spheres. Here, a crossover appears near $\theta_c \sim 24°$. (**b**) For small sphere diameters, the characteristic angle for defect proliferation deviates from its continuum value, with smaller PS spheres triggering the formation of defects at larger polar angles. An idealized prediction for the energy of a single defect provides a rough estimate for the critical angle (blue curve with blue band denoting the uncertainty from the spread in measurements of the defect density). Data for the smallest sphere diameters included only sheets stamped on isolated spheres, not sheets which cover close-packed PS lattices

from ensemble averages over Voronoi tessellations such as that shown in Fig. 3.5b. For each PS sphere diameter D, we identify a characteristic angle at which the number of defects begins to grow significantly (black dashed line in Fig. 3.8a). This analysis leads to the black data in Fig. 3.8b, which shows the characteristic angle as a function of D. This angle approaches a constant value consistent with scale-invariance in the continuum limit of large PS sphere sizes, where the nanoparticle lattice spacing becomes irrelevant. However, we observe an increase in the angle for the smallest PS sphere sizes. This observed variation in the onset of dislocation proliferation suggests that the discrete nature of the lattice becomes important for small D.

If we approximate our sheet as a locally flat, two-dimensional lattice, each dislocation pair costs an elastic energy [110]

$$E_{\text{disloc}} \approx \frac{\mu a^2}{2\pi(1-\nu)} \ln\left(\frac{\ell}{a}\right), \tag{3.5}$$

where Y is the sheet stiffness, ν is the Poisson ratio, ℓ is the final distance between the unbound dislocations, and a is the lattice spacing. We assume the elastic core energy to be small compared to the elastic energy in the deformed sheet, with the understanding that Eq. 3.5 represents a lower bound. Below, we consider $\ell \approx 1/3\sqrt{\rho}$, as illustrated in the inset of Fig. 3.8. Here, ρ is the density of dislocations (so that ρ^{-1} approximates the area of a patch whose elastic deformation is dominated by the dislocation's presence). Note that we expect this elastic energy to be felt predominantly in regions of the material which are not already pinned to the underlying substrate.

In order to find a lower bound for the critical angle at which defects may appear, we compare the dislocation unbinding energy (Eq. 3.5) with the stretching energy for the sheet to conform to a sphere. Using the results from spring network simulations, we equate the stretching energy available in an annulus of width chosen to be $\delta r = a$ with the unbinding energy of Eq. 3.5. This gives the blue solid line in Fig. 3.8 for $\ell = (3\sqrt{\rho})^{-1}$, with the blue band denoting the range of results given the standard deviation of measurements for ρ across sheets on all PS spheres included in the analysis. As seen by the width of the blue band, the prediction is moderately sensitive to the assumed distance that the unbound dislocation travel apart in their creation. We measured the dislocation density, ρ, from the relative frequency of dislocations at $\theta = 0$ in experiments. Despite the approximate nature of the derivation, the prediction lies within our experimental uncertainty for changes in the choice of δr by up to a factor of three, and the agreement in the shape of $\theta_c(D)$ is notable.

3.7.2 Formation of Azimuthal Cracks

Another response to the buildup of strain is to form cracks in a material. This irreversible deformation relieves elastic energy by severing bonds between nanoparticles. We find that, for PS sphere sizes above 210 nm, nanoparticle sheets generally form azimuthal cracks such as those seen in Figs. 3.2c and 3.5.

From a geometric standpoint, projecting an annular strip of inner diameter $\pi D\theta_0$ from a flat disk onto a sphere of diameter D involves less azimuthal compression if the annulus is placed at a polar angle $\theta_{\text{adhere}} > \theta_0$. This fact is reflected in our experiments and simulations, with radial strain building up with increasing polar angle. Once the radial strains are sufficient to rip apart bonds to form azimuthal cracks, we expect that as the next portion of the membrane drapes onto the sphere, it is energetically favorable to adhere to a location further down, at $\theta_{adhere} > \theta_{rip}$. The result is a portion of uncovered PS sphere between θ_{rip} and θ_{adhere}, i.e., an azimuthal crack imprinted on the spherical substrate.

3.8 Formation of Folds at Large Sphere Sizes

For the largest PS sphere sizes, the caps formed by the adhering nanoparticle sheets are large enough that radially oriented folds can be observed (Fig. 3.2d). Such folds provide an alternate mechanism to map circles in the plane to circles on a sphere while minimizing radial tension and azimuthal compression. Localizing elastic energy into folds relieves the stretching in intervening patches. At the same time, because of the very high curvature in one dimension at the fold (which we expect to be comparable to the inverse lattice constant, a^{-1}), the energetic barrier to fold formation is larger than the bending energy by a factor $\sim D^2/a^2$, implying that the cost of having a fold in an annulus of fixed width, δr, does not vary with sphere diameter D. This means that, for sufficiently large D, where the elastic cost of stretching grows higher and higher, fold formation is no longer frozen out (Fig. 3.4).

In previous studies of folding that subjected thin sheets to uniaxial compression or out-of-plane deformation, folds often span the whole system [49, 111, 112], though we note this is not always the case [113]. In our system, the fold terminus occurs at a characteristic polar angle, and the amount of material stored in each fold grows further from the apex of the sphere in order to accommodate the curvature of the underlying substrate (Fig. 3.2d). This type of fold also appears in skirts and other clothing, where it is called a 'dart'.

While we robustly observe pronounced folds on large PS spheres, we find no evidence for smaller-scale wrinkling in the sheets. This can be predicted from the energy scaling (Fig. 3.4): the cost to delaminate from the PS surface exceeds both folding and stretching energies ($E_\gamma > E_f, E_s$).

3.9 Conclusion

In this article, we focused on the ability of preassembled nanoparticle monolayer sheets to conform to a substrate composed of a lattice of larger spheres. With its local Gaussian curvature, G, which can be tuned by varying the sphere diameter, such a substrate serves as a model for arbitrary surface topographies. In the presence of strong pinning to the substrate, the area mismatch between flat ($G = 0$) and spherical ($G > 0$) geometries triggers a competition between different deformation modes of the sheet, including delamination, bending, stretching, fracture, and folding.

Treating the sheets as homogeneous continuum material leads to a scaling picture which is consistent with the general trends of elastic deformation in our system. For comparison with experiments, we extracted the local strain tensor components from images of the sheets, where the nanoparticles served as distance markers. While this analysis was consistent with our general scaling picture, the details of plastic deformation are only captured if the discrete nature of the sheets is taken into account, allowing changes in the number of nearest neighbors for individual particles. By tracking the onset of strain-induced dislocations within the sheets, we are able to explain deviations from the continuum predictions, which are found when the sheets are conformed to substrates with small D, corresponding to regions of large G.

The observed morphologies for the stamped sheets highlight the remarkable ability of nanoparticle monolayers to cope with strain through a combination of elastic and plastic deformations. This material contrasts with other thin sheets such as paper, mylar, or graphene, which lack a similar mechanism for generating particle dislocations. We note that if the material properties of our sheets were tuned by changing the gold nanoparticle size, changing the ligand length, or functionalizing the ligands, a different sequence of morphological regimes could emerge as the substrate sphere size varies.

There is currently much interest in creating functional materials by stacking ultra-thin, essentially 2D layers with different electronic or optical properties [114, 115]. So far, such stacking has been limited to flat substrates, where it is relatively easy to obtain good interfaces between successively deposited layers. In this regard, the ability of nanoparticle sheets to comply and conform opens up new possibilities for creating stacked layers with well-controlled interfaces also on more complex substrate topographies.

Part II
Topological Mechanics in Gyroscopic Metamaterials

Chapter 4
Realization of a Topological Phase Transition in a Gyroscopic Lattice

Topological metamaterials exhibit unusual behaviors at their boundaries, such as unidirectional chiral waves, that are protected by a topological feature of their band structure. The ability to tune such a material through a topological phase transition in real time could enable the use of protected waves for information storage and readout. Here we dynamically tune through a topological phase transition by breaking inversion symmetry in a metamaterial composed of interacting gyroscopes. Through the transition, we track the divergence of the edge modes' localization length and the change in Chern number characterizing the topology of the material's band structure. The work reported in this chapter provides a new axis with which to tune the response of mechanical topological metamaterials.

4.1 Topological Phase Transitions

A central challenge in physics is understanding and controlling the transport of energy and information. Topological materials have proven an exceptional tool for this purpose, since topological excitations pass around impurities and defects and are immune to back-scattering at sharp corners [18, 116]. Furthermore, topological edge modes are robust against weak disorder, such as variations in the pinning energy of each lattice site, in contrast to typical edge waves [16, 117]. This chapter is adapted from [45] with permission.

Mechanical topological insulators represent a rapidly-growing class of materials with topologically-nontrivial phononic band structure [18, 36]. A signature of topological protection is the existence of finite frequency waves around the perimeter of such a material, in a direction determined by the band topology [16, 17, 47, 118, 119]. These edge waves are immune to scattering either into the bulk or in the reverse direction along the edge. Since the physics of topological protection is in many cases agnostic to whether the material is built from classical or quantum

© Springer Nature Switzerland AG 2020
N. Mitchell, *Geometric Control of Fracture and Topological Metamaterials*,
Springer Theses, https://doi.org/10.1007/978-3-030-36361-1_4

components, classical systems in which the individual components are readily accessible offer an appealing arena in which to explore this physics. At the same time, harnessing topological wave behavior for applications motivates real-time control of chiral edge waves, including the ability to tune through a topological phase transition, as has recently been accomplished in quantum materials [120, 121]. Here we present a method for reversibly passing through a topological phase transition in a mechanical metamaterial, which allows us to tune chiral edge modes on and off in real time and the see effects of the transition.

4.2 Experimental Setup

Our system consists of rapidly-spinning gyroscopes hanging from a plate (Fig. 4.1). If displaced from equilibrium, a single gyroscope will precess: its tip moves in a circular orbit about the equilibrium position as a result of the torques from gravity

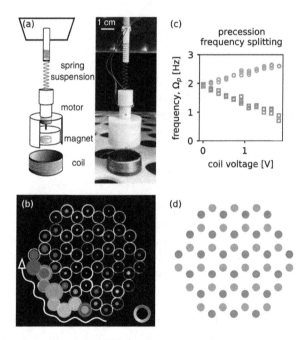

Fig. 4.1 Modulating the magnetic field at each lattice site tunes a metamaterial of gyroscopes suspended from a plate. (**a**) A magnet embedded in each gyro provides an interaction with nearby gyroscopes and with a current-carrying coil. (**b**) A honeycomb network of interacting gyroscopes supports topologically-protected chiral edge waves. Overlaid circles depict the gyroscopes' displacements, colored by the phase of the displacement with respect to the equilibrium positions (see color wheel in the bottom right of panel (**b**)). (**c**) The magnetic field from a coil placed below modulates the precession frequency at each site, raising (the orange circles) or lowering the frequency (the blue squares) depending on the orientation of the current through the coil. (**d**) Modulating the precession frequencies in an alternating pattern breaks inversion symmetry of the lattice

and the spring suspension. To induce repulsive interactions between gyroscopes, we place a magnet in each gyroscope with the dipole moments aligned.

The experimental system differs from that presented in [17, 47] in three ways. The coil beneath each gyroscope is the most important addition, as this drives the transition to the trivial phase by introducing a magnetic field at each site, with alternating field orientation. Secondly, the mass of each gyroscope is increased to ~25 g. The additional mass serves to stabilize the gyroscope position on sites with repulsive magnetic field orientation, as these gyroscopes would otherwise become highly canted for modest magnetic field strength. Lastly, the gyroscopes are driven by a pulse-width-modulated signal, where the duty cycle of the pulse width modulation (PWM) is controlled for each gyroscope individually.

As shown in Fig. 4.2, daisy-chained LED drivers (Texas Instruments TLC5940s) facilitate the parallel PWM control. The drivers use four serial signals to control the

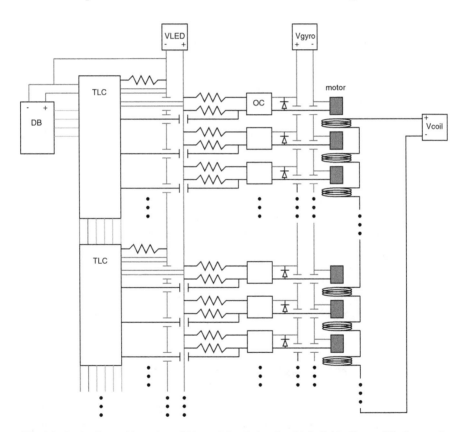

Fig. 4.2 A circuit provides pulse-width-modulated signals with individually tunable duty cycle, allowing for synchronized spinning speeds of the gyroscope motors. An Arduino (DB) provides serial output signals (green) to daisy-chained integrated circuits (TLC5940). The TLCs translate the serial signals into pulse-width-modulated signals sent to parallel optocouplers (OC). The optocouplers then modulate the larger voltage drop across the gyroscope motors. A separate circuit controls the coils beneath each gyroscope

duty cycle of up to 16 gyroscopes each. Optocouplers isolate the drivers from the power supply passing current to the gyroscopes. Each optocoupler passes current to a gyroscope motor when its corresponding TLC pin accepts (sinks) current. The coils, meanwhile, are powered in series in a separate circuit.

4.3   Broken Symmetries in the Honeycomb Lattice

A honeycomb lattice of such gyroscopes behaves as a Chern insulator, exhibiting robust chiral edge waves that pass around corners uninhibited. The phononic spectrum has a band gap, and shaking a boundary site at a frequency in the gap generates a wavepacket that travels clockwise along the edge. Figure 4.1b shows such a wavepacket as seen from below.

The origin of these chiral edge modes is broken time reversal symmetry, which arises from a combination of lattice structure and spinning components [17]. As in [17, 47], an effective time reversal operation both reverses time ($t \rightarrow -t$) and reflects one component of each gyroscope's displacement ($\psi \rightarrow \psi^*$, where $\psi = \delta x + i \delta y$ is the displacement of a gyroscope). Breaking effective time reversal symmetry opens a gap at the Dirac points of the phononic dispersion in a way that endows each band with a nonzero Chern number [122].

An alternative mechanism for opening a gap, however, is to make sites in the unit cell inequivalent [120, 122]. This process breaks inversion symmetry in the honeycomb lattice: the system is no longer invariant under exchange of the two sites in the unit cell. A gap opened by this mechanism is topologically trivial and does not lead to protected edge modes.

If both symmetries are broken, then their relative strength should determine whether the system is topological or trivial, enabling us to tune the system through a phase transition. This would be analogous to a known transition in the Haldane model [122], in which broken inversion symmetry competes with the broken time reversal symmetry. This mechanism for passing through such a topological transition has been used in systems of cold atoms in driven optical lattices [120, 121]. Here, we explore the analogous behavior in a topological mechanical metamaterial.

4.4   Breaking Inversion Symmetry in Experiment

A simple way to break inversion symmetry is to detune the precession frequencies of neighboring gyroscopes, pairwise throughout the system. To do so, we apply a local magnetic field at each site by introducing a coil beneath each gyroscope. For small displacements, the coil's magnetic field provides a force which is parallel or antiparallel to gravity, raising or lowering the its precession frequency: $\Omega_p \to$ $\Omega_p^{A,B} = (1 \pm \Delta_{AB})\Omega_p^0$ (Fig. 4.1c). We then assemble a honeycomb lattice of gyroscopes with alternating coil orientations at each site (Fig. 4.1d). To reduce noise in the precession frequencies, we synchronize all spinning speeds by sending pulse-width-modulated signals to the motors.

We excite a wavepacket in this system, again by shaking a site at the boundary, and simultaneously ramp up the current through the coils. As we pass through a critical current—corresponding to a critical inversion symmetry breaking strength—the excitation delocalizes: the coherent, topologically-protected edge mode transforms into bulk modes, suggesting the presence of a topological phase transition (see bottom panel in Fig. 4.3). While the gradual ramp ($22\,\Omega_p^0$) in Fig. 4.3 allows visual confirmation of the edge mode delocalizing, more rapid ramps likewise halt the edge mode. We note that ramps of less than $\sim 8\,\Omega_p^0$ cause bulk disturbance from the impulsive magnetic torques on gyroscopes that are canted during the ramp.

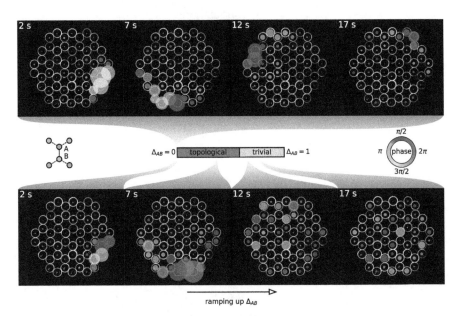

Fig. 4.3 Dynamically ramping up the inversion symmetry breaking quenches a chiral wave. *(Top)* For $\Delta_{AB} = 0$, exciting a mode in the gap by shaking at a frequency in the center of the band gap yields a robust chiral edge wave. *(Bottom)* The same wavepacket is created in the lower panel, but here the inversion symmetry breaking increases over the first 14 s of the experiment. As Δ_{AB} passes through the critical value, the mode delocalizes, and no coherent packet persists in the trivial phase. The system is viewed from below through the coils (white circles). Filled colored circles overlaying each image represent the displacement of the gyroscopes. Each colored circle's radius is proportional to the magnitude of the displacement, while its color represents the displacement's phase (see color wheel)

4.5 Measuring the Topological Phase Transition

To study the transition in more detail, we compute the band structure of magnetic gyroscopes with varying inversion symmetry breaking. As the precession frequency splitting is increased, the band gap closes and reopens (Fig. 4.4a). Beyond the critical value, no edge states connect the two bands: the system is a trivial insulator.

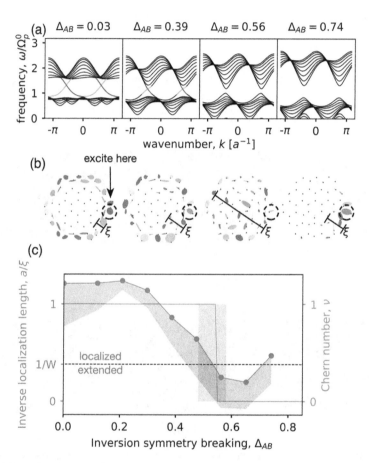

Fig. 4.4 Inversion symmetry breaking drives the topological transition to a trivial insulator, with a divergence in localization length at the transition. (**a**) Band structure for a periodic supercell shows the gap close and reopen; increasing the frequency splitting, Δ_{AB}, eliminates the chiral edge modes localized to the top (purple) and bottom (light green) of the supercell. (**b**) Experimental measurement of the edge states near the center of the gap show the divergence of the localization length. At large Δ_{AB}, exciting the system in the trivial band gap leads to a weak, localized response. (**c**) When the most localized state (connected blue dots) becomes extended such that $\xi \sim W$ (black dashed line), where W is the distance from the center to the system's boundary, the Chern number of the lower band (orange line) changes from +1 to zero. The orange band represents uncertainty in the transition point arising from uncertainty in experimental parameters. All states in the range $[0.9\,\Omega_p^0, 1.1\,\Omega_p^0]$ are included in the blue band, and a is the lattice spacing

Measuring the localization length of edge states in our experiment enables a direct comparison against our model. As the gap narrows, the localization length of the most confined edge state broadens until it is comparable to the system size. By shaking the system at a frequency that slowly sweeps through the gap and tracking the gyroscopes' displacements, we obtain the eigenstates of the system (the eigenvectors of the Fourier transform). Figure 4.4b and c show the results of this measurement, considering only the most localized states near the center of the gap. As the localization length of the most localized state (blue curve in Fig. 4.4c) grows to the scale of the system size, W, the bands touch and reopen without chiral edge modes. This feature confirms the likely presence of a topological phase transition.

To predict the topological phase transition theoretically, we compute the Chern number of the system's bands, which is encoded in the spectrum of the dynamical matrix. To linear order, the displacement of a fast-spinning gyroscope, $\psi \equiv \delta x + i \delta y$, obeys Newton's second law as

$$i\frac{d\psi_p}{dt} = \Omega_p \psi + \frac{1}{2}\sum_q \left[\left(\Omega_{pp}^+ \psi_p + \Omega_{pq}^+ \psi_q \right) \right.$$
$$\left. + e^{2i\theta_{pq}} \left(\Omega_{pp}^- \psi_p^* + \Omega_{pq}^- \psi_q^* \right) \right], \quad (4.1)$$

where the sum is over nearby gyroscopes, $\Omega_{pq}^\pm \equiv -\frac{\ell^2}{I\omega_0}\left(\partial F_{p\parallel}/\partial x_{q\parallel} \pm \partial F_{p\perp}/\partial x_{q\perp}\right)$ is the characteristic interaction frequency between gyroscopes p and q, $\Omega_p \equiv (mg + F^{\text{suspension}} + F_z^{\text{coil}})\ell/I\omega_0 = (1 + \Delta_{AB})\Omega_p^0$ is the precession frequency in the absence of other gyroscopes, and θ_{pq} is the angle of the bond connecting gyroscope p to gyroscope q, taken with respect to a fixed global axis. The interaction strengths, $\Omega_{pq}^\pm$, scale with the quantity $\Omega_k \equiv \ell^2 k_m/I\omega_0$, where k_m is the effective spring constant for the magnetic interaction, ω_0 is the spinning speed of the gyroscope, and $\Omega_{pq}^\pm$ depend nonlinearly on the lattice spacing. As Eq. 4.1 resembles the Schrödinger equation, we write the equation of motion for the entire system as

$$i\frac{d\boldsymbol{\psi}}{dt} = \mathbf{D}\boldsymbol{\psi}. \quad (4.2)$$

The precession frequency plays the role of the on-site potential, so that the coil's magnetic field detunes the diagonal terms of the dynamical matrix, $\mathbf{D}$.

A nonzero Chern number signals the existence of topologically-protected chiral edge modes. Computing the Chern number of the magnetic system as

$$C_j dx \wedge dy = \frac{i}{2\pi} \int d^2k \, \text{Tr}\left(dP_j \wedge P_j dP_j\right), \quad (4.3)$$

where the projector $P_j \equiv |u_j\rangle\langle u_j|$ maps states in band j to themselves and maps other states to zero, we see the Chern number of the lower band change from 1 to 0 when the localization length reaches the system size, $\xi \sim W$ (Fig. 4.4c).

4.6   Competing Broken Symmetries

Our approach enables us to tune through a phase transition dynamically by
pitting inversion symmetry against time reversal symmetry breaking, adding a new
axis of versatility for topological mechanical metamaterials. We illustrate this by
computing a larger phase diagram for the gyroscopic system in which time reversal
symmetry breaking and inversion symmetry breaking are both varied. To explore
their interplay, we combine the transition observed here with another topological
phase transition discussed in [17], which exploits the dependence of the band
topology on the geometry of the lattice. By globally deforming the honeycomb
lattice through a brick-layer lattice, the Chern number of the lower band transitions
from 1 to 0 to -1 (Fig. 4.5). The transition occurs when bond angles in the network
are precisely multiples of $\pi/2$, at which point effective time reversal symmetry is
restored and the gap closes. Continuing the deformation into a bowtie configuration
inverts the sign of the symmetry-breaking term, reopening the gap, but each band's
Chern number flips sign.

In Fig. 4.5, we allow inversion symmetry breaking and lattice deformation to
compete, giving rise to systems with clockwise edge modes (red), counterclockwise
edge modes (blue), and no chiral edge modes (white). When the time reversal

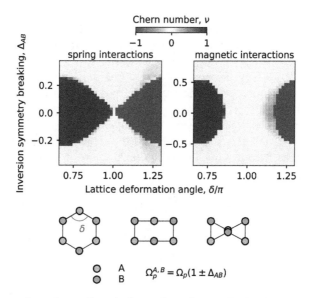

Fig. 4.5 Our experiment is one slice of a larger phase diagram. Tuning the lattice geometry to a
bricklayer lattice restores effective time reversal symmetry, and continuing to deform into a bowtie
inverts the sign of the symmetry-breaking term. The left panel shows the topological phase diagram
for the case of spring couplings (no restoring force in the transverse direction, to first order), with
$\Omega_p^0 = \Omega^+ = \Omega^-$. The right panel shows the Chern number of the lower band for the case of
magnetic interactions, with $\Omega_k/\Omega_p^0 = 0.67$, as in the experiments. The two are similar, though
the topologically nontrivial phases (red and blue) do not meet at a point in the case of magnetic
interactions

symmetry breaking is weakened ($\delta \to \pi$), the required $|\Delta_{AB}|$ to drive the system to a trivial insulator diminishes. This phase diagram highlights the similarity of the gyroscopic system to the Haldane model [122]. This can be understood by taking the limit in which the precession frequency is much faster than the interaction strength ($\Omega_p^0 \gg \Omega_k$): the spring-coupled gyroscope system maps to the Haldane model [17].

The phase diagram for magnetic interactions, while similar, possesses an area of trivial insulator between the topologically nontrivial phases. Unlike spring-like potentials, magnets exhibit an anti-restoring response to perpendicular displacements ($\Omega^+ \neq \Omega^-$). These interactions can introduce an indirect band gap that closes before the lattice reaches a bricklayer geometry. Figure 4.6 shows the band structure for fully periodic lattices of spring-coupled and magnet-coupled gyroscopes as the lattice deformation angle, δ, increases. In the magnetic case, the frequency gap

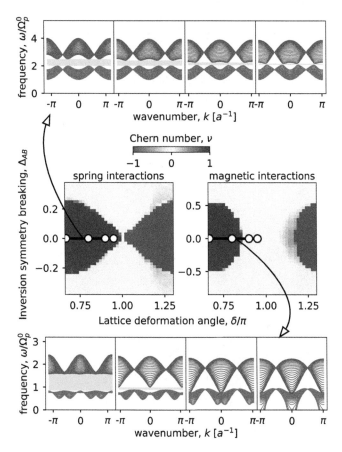

Fig. 4.6 An indirect band gap closing is responsible for the difference between the spring-coupled and magnetically-coupled phase diagrams. Representative band structures computed for a fully periodic lattice show the approach to the topologically trivial phase as the lattice deformation angle, δ, increases. With spring interactions, the topological phase transition occurs as the band gap (yellow) closes at the Dirac points (top panel sequence). In contrast, with magnetic interactions, the gap closes before the two bands touch at any single wavevector

vanishes before the two bands touch at any single wavevector, in contrast to the spring-coupled case. The result is a trivial insulator phase of finite extent separating the phases with $\nu = +1$ and $\nu = -1$ for the lower band (right panel in Fig. 4.5). The extent of the separation depends on the interaction strength $\Omega^{\pm}/\Omega_p^0$ and the spacing between gyroscopes relative to their pendulum length.

4.7   Conclusion

We have demonstrated a non-destructive mechanism for dynamically tuning a mechanical Chern insulator through a topological phase transition. We characterized the transition by measuring the delocalization of edge modes in gap and the corresponding change in Chern number to zero at the transition, and we established this 1D transition's context within a 2D phase space for mechanical gyroscopic Chern insulators. This design enables a mechanism for constructing topological gates able to direct the flow of energy in chiral modes, offering potential applications in classical information storage and readout [123].

Chapter 5
Tunable Band Topology in Gyroscopic Lattices

As we have discussed, gyroscopic metamaterials—mechanical structures composed of interacting spinning tops—support one-way topological edge waves. In these structures, the time reversal symmetry breaking that enables their topological behavior emerges directly from the lattice geometry. Here we show that variations in the lattice geometry can therefore give rise to more complex band topology than has been previously described. A 'spindle' lattice (or truncated hexagonal tiling) of gyroscopes possesses both clockwise and counterclockwise edge modes distributed across several band gaps. Tuning the interaction strength or twisting the lattice structure along a Guest mode opens and closes these gaps and yields bands with Chern numbers of $|C| > 1$ without introducing next-nearest-neighbor interactions or staggered potentials. A deformable honeycomb structure provides a simple model for understanding the role of lattice geometry in constraining the effects of time reversal symmetry and inversion symmetry breaking. Lastly, we find that topological band structure generically arises in gyroscopic networks, and a simple protocol generates lattices with topological excitations. This chapter is adapted from [46] with permission.

5.1 Gyroscopic Lattices

Materials with nontrivial band topology have captured the attention of condensed matter scientists since their discovery in electronic systems [42]. Since then, the concept of topological order has found its way to a plethora of physical systems, from electronic to photonic, acoustic, and even mechanical systems [16, 17, 36, 45, 47, 117–119, 124–128]. When topologically nontrivial, all these systems exhibit excitations confined to their surface that propagate unidirectionally without backscattering and are robust to disorder. These features are both fundamentally intriguing and form the basis for technological applications of topological materials.

© Springer Nature Switzerland AG 2020
N. Mitchell, *Geometric Control of Fracture and Topological Metamaterials*,
Springer Theses, https://doi.org/10.1007/978-3-030-36361-1_5

Here, we focus on unidirectional edge modes in structures composed of coupled spinning objects [17, 45, 47, 119, 129, 130]. In particular, we focus we focus on the topological properties arising from the collective motion of lattices of gyroscopes—namely, how their phononic band structure encodes a nonzero Chern number. When the band structure is topologically nontrivial, the gyroscopic system supports unidirectional waves on its boundary. These edge waves are distinct from a range of other non-reciprocal properties that emerge in angular momentum-biased systems because of their topological origin [39, 131].

The minimal requirements for such a Chern insulator are the presence of a band gap and broken time reversal symmetry. In the electronic case, time reversal symmetry breaking arises from the presence of magnetic fields [122]. As we will see, the analogous mechanism in gyroscopic lattices is the lattice geometry itself: the mere presence of spinning components is not sufficient to generate the effects enabling chiral edge modes.

In this chapter, we go beyond simple geometries and find the flexibility to design lattices with desired band gaps and desired topology. In particular, we examine tunable lattices with Chern numbers $|C| > 1$ as well as multiple gaps with edge modes of opposite chirality. We examine the effects of competing time reversal symmetry breaking with inversion symmetry breaking, and demonstrate a design strategy to achieve band topology in lattices with seemingly arbitrary unit cells.

5.2 The Equations of Motion

A simple realization of gyroscopic metamaterials is a collection of coupled gyroscopes which hang from a pivot point and spin rapidly enough for their angular momentum to lie approximately along the primary axis, as shown in Fig. 5.1a. Under these conditions, the free tip of a gyroscope moves when a torque, $\vec{\tau}$, acts about the pivot point according to:

$$\vec{\tau} \approx I\omega_0 \partial_t \hat{n} = \vec{\ell}_f \times \vec{F} \tag{5.1}$$

where I is the principal moment of inertia, ω_0 is the spinning speed, $\hat{n}$ is a unit vector pointing from the pivot point toward the center of mass, and $\vec{\ell}_f$ is the vector from the pivot point to the point acted upon by force, $\vec{F}$.

Considering small displacements of each gyroscope allows a linearized description. Denoting the displacement from the equilibrium position in the plane as $\psi = x + iy$, the equation of motion for a single gyroscope under the influence of gravity becomes:

$$i\partial_t \psi = \frac{mg\ell_{cm}}{I\omega_0} \psi. \tag{5.2}$$

Note that throughout this chapter, without loss of generality, we choose the angular momentum vector of a hanging gyroscope to point down, from the pivot

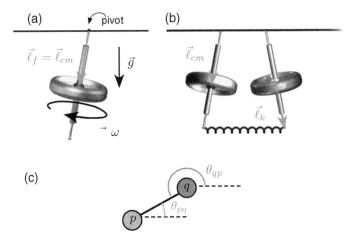

Fig. 5.1 A spring-coupled gyroscopic metamaterial is composed of spinning gyroscopes that hang from a pinned pivot point. (**a**) $\vec{\ell}_f$ is the vector from the pivot point to where a force acts. When the only force is gravity, $\vec{\ell}_f = \vec{\ell}_{cm}$. (**b**) Gyroscopes in the metamaterial are coupled to their neighbors in the lattice via a spring which is attached to the free end. (**c**) The linearized equation of motion for our system relates the displacements via angles between bonds and the local gyroscope's local x-axis (indicated by dotted lines in this view from below)

point towards the center of mass. Noting the similarity between Eq. 5.2 and the Schrödinger equation for a quantum particle, we use the same notion of time reversal symmetry as is used in quantum mechanics, namely $\psi \rightarrow \psi^*$ and $t \rightarrow -t$. While $\psi \rightarrow \psi^*$ corresponds to a reversal of momentum for a quantum particle, in the context of gyroscopes, $\psi \rightarrow \psi^*$ carries out a reflection of the gyroscope's displacement about a horizontal axis passing through its pivot point. Performing this operation on the equation above, we find that Eq. 5.2 is time reversal symmetric. Thus, a spinning top precessing under the influence of gravity does not break this notion of time reversal symmetry.

Introducing interactions, however, allows the structure to break time reversal symmetry. The simplest setting to see this is a network of gyroscopes coupled by linear springs. For small displacements, the forces exerted on one gyroscope by another are proportional to the component of the net displacement along the line connecting them. For a given pair of gyroscopes p and q, it is convenient to extract the component of the net displacement $\psi_p - \psi_q$ along the bond by rotating the system to the local x-axis of p, taking the real part of expression, and then rotating back. The resulting force in complex form is given by:

$$F_{pq} = -k_0 e^{i\theta_{pq}} \mathrm{Re}[e^{-i\theta_{pq}} (\psi_p - \psi_q)] \tag{5.3}$$

$$= -\frac{k_0 e^{i\theta_{pq}}}{2} \left[e^{-i\theta_{pq}} (\psi_p - \psi_q) + e^{i\theta_{pq}} (\psi_p^* - \psi_q^*) \right],$$

where k_0 is the spring constant of the bond. Using this result, the equation of motion for two gyroscopes can then be written as:

$$i \partial_t \psi_p = \Omega_g \psi_p + \frac{\Omega_k}{2} \left[(\psi_p - \psi_q) + e^{2i\theta_{pq}} \left(\psi_p^* - \psi_q^* \right) \right]. \tag{5.4}$$

where $\Omega_k = k_0 \ell_k^2 / (I\omega_0)$ and $\Omega_g = mg\ell_{cm} / (I\omega_0)$. We define the time reversal operation as $\psi_p^{TR}(t) = \psi_p^*(-t)$. By taking the complex conjugate of $\partial_t \psi_p$ and rewriting in terms of ψ_p^{TR}, we see that the equations of motion are changed only by $e^{i2\theta} \to e^{-i2\theta}$. Therefore, we see that time reversal symmetry is preserved under reflections that are parallel or perpendicular to the bond [17].

The full equation of motion for a hanging gyroscope with more than one neighbor can be similarly expressed:

$$i \partial_t \psi_p = \Omega_g \psi_p + \frac{\Omega_k}{2} \sum_q^{n.n.} \left[(\psi_p - \psi_q) + e^{2i\theta_{pq}} \left(\psi_p^* - \psi_q^* \right) \right]. \tag{5.5}$$

As before, time reversal symmetry is only preserved if all bonds are either parallel or perpendicular to each other since in this case, a coordinate system can be chosen so that bonds lie along the x and y axes, constraining the prefactor $e^{2i\theta_{pq}}$ to be real for all bonds in the network.

To date, the only ordered lattices that have been considered in this framework are the honeycomb lattice and simple distortions thereof [17, 45]. A slightly different manifestation of gyroscopic metamaterials considered in [119] found that by including staggered sublattice precession frequencies and bond strengths, time reversal symmetry could be effectively broken in lattices with square and honeycomb symmetries.

5.3 Twisted Spindle Lattice

To demonstrate the considerable flexibility of gyroscopic metamaterials, we begin by considering the twisted spindle lattice shown in Fig. 5.2. This structure shares features of both the honeycomb lattice and the kagome lattice. As Fig. 5.2a shows, shrinking the blue triangles to a single site—while increasing the strength of blue bonds—deforms the spindle lattice into the honeycomb lattice. Conversely, taking the length of the red bonds that connect triads of gyroscopes to zero—while increasing their strength—deforms the spindle lattice into a kagome configuration. As shown in Fig. 5.2b, the spindle lattice also supports a Guest mode—a global elastic distortion that costs no energy—in which each triad of gyroscopes rotates locally.

In the limiting case of the honeycomb lattice, which has two sites per unit cell, we find a single gap with clockwise topologically-protected edge modes [17]. By

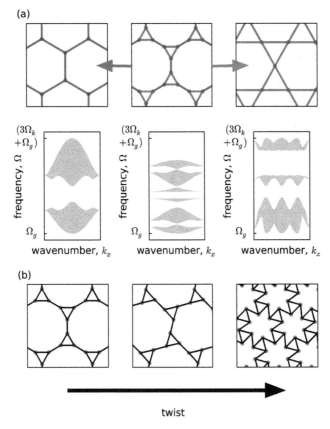

Fig. 5.2 The spindle lattice shares features of both the honeycomb and kagome lattices, while supporting a Guest mode in which each triad of gyroscopes is locally rotated. (**a**) Taking the size of the red triangles in the spindle lattice to zero returns a honeycomb configuration, while taking the length of the blue bonds connecting each red triangle of gyroscopes to zero transforms the spindle lattice into the kagome configuration. The associated band structures are shown below each of the three configurations. (**b**) Locally twisting the triangles of a spindle lattice preserves bond lengths while globally deforming the lattice

contrast, in the kagome lattice, with three sites per unit cell, there are two gaps, which each support a counterclockwise topological mode. In the intermediate case of the undeformed spindle lattice, which has six sites per unit cell, we generically find five band gaps. Most of these gaps possess chiral edge modes, and a given configuration can host both clockwise and counterclockwise modes. As we show below, locally twisting this structure (as in Fig. 5.2b) or varying the bond strengths, Ω_k, relative to the pinning strength, Ω_g, opens and closes edge-mode-carrying gaps.

Shaking a gyroscope on the boundary of this network at a frequency in the lowest band gap generates a clockwise wavepacket confined to the edge of the sample which is robust to disorder in the gravitational precession frequencies or bond strengths and does not scatter at sharp corners or defects (Fig. 5.3a). Shaking

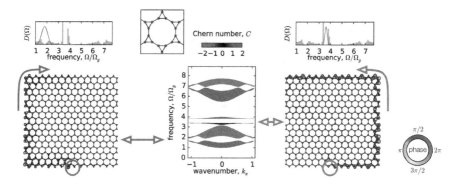

Fig. 5.3 The gyroscopic spindle lattice contains chiral edge modes of either chirality as well as a band with Chern number of $|C| > 1$. Direct simulation of Eq. 5.5 reveal clockwise (left) and counterclockwise (right) edge modes in the same structure when shaken at different frequencies ($\Omega = 1.8\,\Omega_g$ (left) and $3.62\,\Omega_g$ (right)). The displacement of each gyroscope is represented as a circle with a radius proportional to the displacement's magnitude. The color of each circle represents the phase of the displacement, as depicted in the color wheel on the bottom right. Computing the Chern numbers for each band confirms the topological origin of the chiral edge modes, as shown by the colored band structure in the middle panel. A single gyroscope on the edge is shaken at a fixed frequency with an amplitude varying in time; the spectrum of the excitation is indicated by red (left) and blue (right) curves overlaying the density of states, $D(\Omega)$. The density of states, shown above each lattice, is given for the case with periodic boundary conditions. For these simulations, the interaction strength was set to be $\Omega_k = 3\Omega_g$

at a frequency in the middle band gap, however, generates counterclockwise edge waves, allowing a single lattice structure to conduct protected edge waves with a chirality determined by frequency (Fig. 5.3b). We compute the Chern number for each band via [132]

$$C_j dx \wedge dy = \frac{i}{2\pi} \int d^2k \, \text{Tr}[dP_j \wedge P_j dP_j], \qquad (5.6)$$

where P_j is the projection matrix defined using a symplectic inner product between states (See Appendix C for details), and where $\wedge$ is the wedge product. We find that the Chern number is equal to the number of chiral edge modes, which suggests the same bulk-boundary correspondence for these systems as in electronic Chern insulators [133].

The topological band structure of the gyroscopic spindle lattice offers additional axes of tunability through varying the interaction strength (i.e. the bond stiffness in the case of springs) and by performing bond-length-preserving deformations on the lattice. For gyroscopic lattices with uniform interaction strengths (i.e. equal spring constants throughout), we can tune the the ratio of interaction frequency to gravitational precession frequency, Ω_k / Ω_g. This operation can close band gaps in addition to changing the frequency range of the band structure. In the case of the spindle lattice, this provides a tuning knob that changes the topology of the band structure. Simply increasing the interaction strength relative to the gravitational

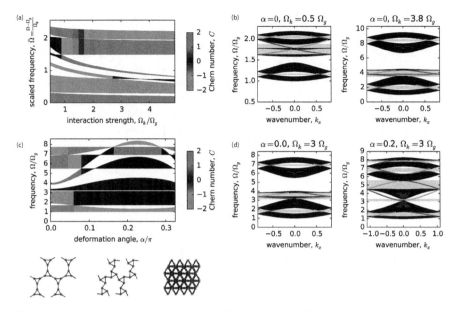

Fig. 5.4 Phononic band structures for the spindle and twisted spindle lattices show opening and closing of gaps with topological edge states. (**a**) Simply increasing the interaction strength enables the closing and opening of band gaps, creating and annihilating protected chiral edge modes. (**b**) Band gaps with chiral edge modes are highlighted in green for two different interaction strengths. (**c–d**) As the structure is twisted through a bond-length-preserving deformation, three of the five gaps close and reopen, leading to three or four gaps with chiral edge modes, depending on the value of the twist deformation angle, α. In panels (**c**) and (**d**), we set $\Omega_k = 3\Omega_g$

precession frequency closes and reopens gaps and changes the Chern numbers of bands, as shown in Fig. 5.4a. We note that this feature was absent in the gyroscopic honeycomb lattice previously studied [17, 45], whose topology was unaffected by changes in Ω_k and Ω_g. This allowed the topology to be continuously connected to the electronic Haldane model, unlike in the spindle lattice.

Twisting the spindle lattice through a Guest mode, as shown in Fig. 5.2b, also provides a tuning knob. Globally deforming the lattice closes and reopens the lowest and highest band gaps, allowing for several distinct configurations of multiple gaps supporting protected chiral edge modes, as shown in Fig. 5.4c. As the twist angle grows, there are five values for which a pair of bands touch and reopen, flipping the chirality of the modes in that gap or imparting chiral modes to a gap which previously had none.

What determines the chirality of edge modes? Unlike in the topological zero-energy modes recently found in Maxwell lattices [36], here the coordination number alone does not play a central role in determining band topology. If Chern numbers were determined purely by the number of nearest neighbors, we would expect, for instance, that the spindle and honeycomb lattices would have similar edge modes: both have a coordination number of $z = 3$. However, the spindle supports edge modes of either chirality. Furthermore, the spindle lattice's rich band structure depends not only on geometry, but also bond strengths (Fig. 5.3a). We conclude that simple, local aspects of the lattice such as coordination number and mean bond angle do not singlehandedly determine the band structure.

From a design perspective, the two simple tuning parameters of angle and inter-action strength are sufficient to cover a broad range of topological phenomenology without introducing staggered interaction strengths, including edge modes with either chirality, the opening and closing of gaps, and bands with Chern number ± 1 and ± 2. These behaviors demonstrate the versatility of gyroscopic metamaterials.

5.4 Time Reversal Symmetry and Topological Bandgaps

All configurations shown so far break time reversal symmetry, which is a necessary ingredient for band topology in Chern insulators [17, 119]. This is not necessarily true for all gyroscopic lattices. For example, as illustrated in Fig. 5.5, a honeycomb lattice can undergo a bond-length-preserving deformation to a configuration in which all bond angles are multiples of $\pi/2$ (for $\delta = \pi$). In such a configuration, time reversal symmetry is restored and therefore band topology disappears. Further changing the value of δ past π causes the band topology to reappear, but with opposite sign. In Fig. 5.5, we extend this analysis to the entire phase-space of periodic, bond-length-preserving deformations by introducing an additional angle, ϕ. This allows us to explore the question of whether time reversal symmetry breaking is sufficient to generate band topology in gyroscopic metamaterials.

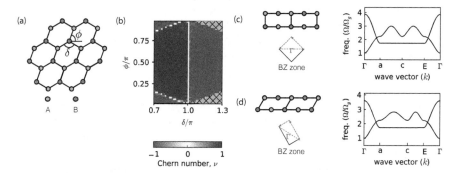

Fig. 5.5 Band gaps and topology in the deformed honeycomb lattice. (**a**) The angles ϕ and δ control the deformation of the honeycomb lattice. (**b**) The ϕ-δ phase diagram shows that the Chern number of the lower band changes when straight lines of bonds appear in the lattice, which occurs on the white diagonal lines in the left corners and on the white vertical line at $\delta = \pi$. (**c**) The bricklayer configuration ($\delta = \pi$) band structure is plotted along paths in the Brillouin zone. The gap is closed at two Dirac points. (**d**) No band gap opens in the canted bricklayer ($\delta = \pi, \phi \neq \pi/2$), even though time reversal symmetry is broken in this configuration

Figure 5.5b shows the topological phase diagram corresponding to general deformations of the honeycomb lattice, characterized by angles ϕ and δ. Red (blue) regions indicate to a Chern number of 1 (-1) for the lower band and, correspondingly, clockwise (counter-clockwise) propagating modes in the gap. For $\phi = \pi/2$ and $\delta = \pi$, the network is arranged in a bricklayer configuration (Fig. 5.5c). Varying either ϕ or δ from this point breaks time reversal symmetry. However, only changes in δ imbue nontrivial band topology, as illustrated by the white line in Fig. 5.5b for $\delta = \pi$ [17]. The fact that changes in ϕ break time reversal symmetry without opening a gap demonstrates that broken time reversal symmetry does not inevitably lead to either band gaps or nontrivial band topology.

This behavior warrants further investigation. During the deformation of the honeycomb into the bricklayer geometry, the band gap closes and the two Dirac points in the spectrum touch at a point. Surprisingly, these Dirac points are preserved even in the canted bricklayer configuration, as shown in Fig. 5.5d, despite the fact that shearing the bricklayer configuration breaks time reversal symmetry by creating acute and obtuse bond angles (see Eq. 5.5).

This protection of the Dirac cones arises due to a subtle pseudo-reflection symmetry. The symmetry consists of reflecting the positions of gyroscopes about the x axis and shearing their relative positions such that the tilt angle ϕ is invariant, while leaving the gyroscopes' displacements unchanged. This pseudo-reflection is a symmetry of the equations of motion, and thus of the normal modes. This symmetry leads to the existence of a special line of modes in momentum space. Along this line, modes that are symmetric and antisymmetric under the symmetry operation decouple and cannot hybridize at their band crossing. Thus, a pseudo-reflection symmetry stabilizes the Dirac points against acquiring gaps, which would otherwise be unstable to time-reversal-symmetry-breaking perturbations—analogous to the effect other discrete symmetries in electronic systems [134]. The pseudo-reflection symmetry also explains the vanishing Chern number for all values of ϕ at $\delta = \pi$ seen in Fig. 5.5b, on account of the Berry curvature being odd under the action of the symmetry. More broadly, this protection underscores of the interplay between lattice geometry and the topological character of the band structure.

5.5 Competing Symmetries in Topological Gyroscopic Systems

Breaking inversion symmetry is the canonical mechanism for opening gaps in the phonon spectra of mass-and-spring lattices [135]. This is also true in other systems, such as electronic materials. This gap opening mechanism can be made to compete with broken time reversal symmetry to close and reopen gaps and eliminate protected chiral edge modes. To study an analogous effect in gyroscopic lattices, we detune the yellow and blue sublattice sites in Fig. 5.6 by modulating their on-site gravitational precession frequencies: $\Omega_{gA,B} = (1 \pm \Delta)\Omega_g$ (see also [45]).

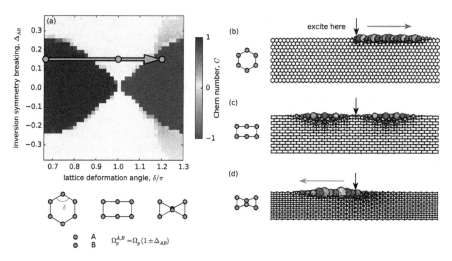

Fig. 5.6 Inversion and time reversal symmetries compete in a gyroscopic lattice. (**a**) The phase diagram for a deformed honeycomb lattice (without shear, so that $\phi = \pi/2$) with varying Ω_g values on sites A and B shows an interplay between inversion and time reversal symmetries. (**b**) In a simulation of the honeycomb lattice with inversion symmetry breaking $\Delta_{AB} = 0.15$, driving a gyroscope on the edge at a gap frequency results in a clockwise wavepacket. (**c**) When the lattice is deformed to a bricklayer geometry, the Chern number vanishes. This configuration is gapped due to the inversion symmetry breaking ($\Delta_{AB} = 0.15$). The gap contains modes which are localized on the edge, but these unprotected edge waves propagate in both directions and are not robust against disorder. (**d**) In the bowtie geometry, edge modes propagate counterclockwise, as predicted by the calculations shown in (**a**)

Figure 5.6a shows the phase diagram that results from varying δ and lattice pinning frequencies, $\Omega_{gA,B}$. When the unit cell's two sites are equivalent ($\Delta = 0$), the Chern number of the system changes only when the gap closes at the bricklayer transition. For $\Delta \neq 0$, however, a third, topologically trivial region appears. In this case, the band structure is gapped, yet displays no chiral edge modes.

The behavior of excitations confirms the Chern number calculations in all three regions, as indicated in Fig. 5.6b–d. In Fig. 5.6c, excitations propagate along the edge in both directions. The Chern number is zero, and these edge waves are not topologically protected: they backscatter at sharp corners or in the presence of disorder. The result shown in Fig. 5.6a displays a strong resemblance to Haldane's phase diagram: sites must have similar pinning strengths for the lattice to support topological states.

While varying precession frequencies is an effective way of breaking inversion symmetry, it is not the only one. An alternative way is to alter the coordination number between sites—i.e. the number of bonds that are linked to each gyroscope. For example, unlike the lattices considered so far, the α–$(ET)_2I_3$ lattice shown in Fig. 5.7 contains sites of coordination number $z = 4$ (for the A and B sites) and $z = 2$ (for the C and D sites). When all gravitational precession frequencies are equal, the lattice displays no topological excitations (top right corner of Fig. 5.7b).

Fig. 5.7 Coordination
number and topological
phases. (**a**) A lattice with four
lattice sites per unit cell,
where sites A and B have two
neighbors and sites C and D
have two. (**b**) The topological
phase diagram for varying the
gravitational precession
frequencies on sites A and B
shows that because of the
different coordination
numbers for the lattice sites,
the band structure is trivial
when
$\Omega_{gA} = \Omega_{gB} = \Omega_{gC} = \Omega_{gD}$.
(**c**) The band structure in the
nontrivial phase for a strip
which is infinite along y and
120 unit cells wide in x

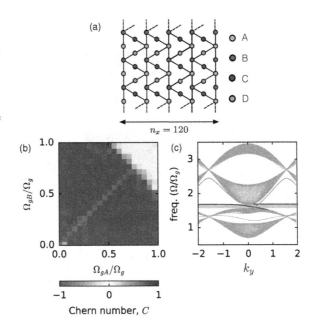

As seen in the first term of Eq. 5.5, contributions to on-site pinning—ie. terms in which $\dot{\psi}_p$ depends on ψ_p—come not only from gravitation precession terms (Ω_g), but also from coupling to adjacent sites. For lattices with unequal coordination at different sites, balancing the full 'site pinning frequency', Ω_p, for each site can be used to enhance or remedy the effects of site inequivalence:

$$\Omega_p \equiv z \frac{\Omega_k}{2} + \Omega_g. \tag{5.7}$$

We can test if site pinning inequivalence is the mechanism preventing the $\alpha-$ $(ET)_2I_3$ lattice from having gaps. Indeed, reducing the precession frequencies of the sites with higher coordination numbers enables a band gap with chiral edge modes (Fig. 5.7b). This provides another example of the inextricable connection between lattice geometry and topological order in gyroscopic lattices.

5.6 Towards Topological Design

We have seen that both time-reversal symmetry and site equivalence are tied to lattice geometry and connectivity. Turning now toward engineering new topological gyroscopic lattices, we can summarize the principles of the previous sections as follows:

1. Breaking time reversal symmetry via bond angles is a necessary, but not sufficient condition for creating a lattice with a non-trivial band topology.

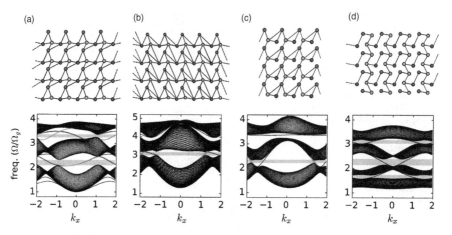

Fig. 5.8 Examples of topological lattices created balancing coordination by varying on-site precession frequencies. Each lattice is generated by placing triangulated points in a square unit cell, then deleting some bonds randomly. (**a**) An example of a deformed kagome lattice structure exhibits two topological gaps. (**b**) A mechanically stable lattice with one topological gap (upper gap) demonstrates that gyroscopic lattices need not be undercoordinated to be Chern insulators. The propagation of edge modes is in the same direction as the kagome lattice. (**c**) A 3-site-per-unit-cell lattice structure with one topological gap (lower gap). The propagation of edge modes is in the same direction as the honeycomb lattice. (**d**) An example of a 4-site-per-unit-cell lattice structure with three topological gaps. The propagation of edge modes is in the same direction as the kagome lattice for all three gaps

2. A competition between time reversal symmetry and site equivalence determines whether or not a lattice can have topological modes. Lattice connectivity is relevant for determining the effective on-site precession frequencies to achieve equivalence.

Using these two principles, one can construct topological metamaterials beginning with an arbitrary unit cell and subsequently balancing pinning frequencies according to Eq. 5.7. This procedure can generate lattices with desired properties—such as multiple bandgaps or mechanical stability. Figure 5.8 shows several examples.

One example of a mechanically stable lattice with non-vanishing Chern number is shown in Fig. 5.8b. Although all previous lattices in this thesis have been mechanically unstable ($\bar{z} \leq 4$), the lattice in Fig. 5.8b shows that this is not necessary for band topology to arise. Sublattices A (yellow) and C (red) have five bonds each, while sublattice B (blue) has four. We expect that topological modes will arise when the total pinning at each site are approximately equal, which would occur for $\Omega_B > \Omega_{A,C}$. Numerics agree with this prediction.

The results demonstrated in this section show that topology is not specific to one family of lattices in gyroscopic networks and is in fact ubiquitous. Many topological lattices can be created using only simple principles—opening a myriad possibilities for material design.

5.7 Conclusion

In this chapter, we explored the interplay between lattice geometry and topological order in gyroscopic lattices—including the effects of broken time reversal symmetry and site equivalence. Along the way, we found examples of lattices with multiple band gaps containing edge modes of either chirality in the same structure and Chern numbers $|C| > 1$. We then identified general principles which are helpful in designing lattices with desired topological band structures. Building on our observations, we used a simple prescription that yields mechanically stable topological gyroscopic lattices and lattices with multiple band gaps. The ubiquity of band topology in gyroscopic metamaterials provides a broad palette with which to design topological behaviors in elastic structures. Further study could investigate the interplay between band topology and nonlinear excitations in gyroscopic networks or interspersing both clockwise and counterclockwise spinning sites.

Chapter 6
Topological Insulators Constructed from Random Point Sets

The discovery that the band structure of electronic insulators may be topologically non-trivial has revealed distinct phases of electronic matter with novel properties [122, 136]. Recently, mechanical lattices have been found to have similarly rich structure in their phononic excitations [36, 118], giving rise to protected unidirectional edge modes [16, 17, 119]. In all these cases, however, as well as in other topological metamaterials [36, 137], the underlying structure was finely tuned, be it through periodicity, quasi-periodicity or isostaticity. Here we show that amorphous Chern insulators can be readily constructed from arbitrary underlying structures, including hyperuniform, jammed, quasi-crystalline, and uniformly random point sets. While our findings apply to mechanical and electronic systems alike, we focus on networks of interacting gyroscopes as a model system. Local decorations control the topology of the vibrational spectrum, endowing amorphous structures with protected edge modes—with a chirality of choice. Using a real-space generalization of the Chern number, we investigate the topology of our structures numerically, analytically and experimentally. The robustness of our approach enables the topological design and self-assembly of non-crystalline topological metamaterials on the micro and macro scale. This chapter is adapted from [47] with permission.

6.1 Gyroscopic Metamaterials as a Model System

Condensed matter science has traditionally focused on systems with underlying spatial order, as many natural systems spontaneously aggregate into crystals. The behavior of amorphous materials, such as glasses, has remained more challenging [138]. In particular, our understanding of common concepts such as bandgaps and topological behavior in amorphous materials is still in its infancy when compared to crystalline counterparts. This is not only a fundamental problem; advances in modern engineering, both of metamaterials and of quantum systems,

© Springer Nature Switzerland AG 2020
N. Mitchell, *Geometric Control of Fracture and Topological Metamaterials*,
Springer Theses, https://doi.org/10.1007/978-3-030-36361-1_6

has opened the door for the creation of materials with arbitrary structure, including amorphous materials. This prompts a search for principles that can apply to a wide range of amorphous systems, from interacting atoms to mechanical metamaterials.

In the exploration of topological insulators, conceptual advances have proven to carry across between disparate physical realizations, from quantum systems [120], to photonic waveguides [117], to acoustical resonators [129, 139], to hinged or geared mechanical structures [36, 140]. One promising model system is a class of mechanical insulators consisting of gyroscopes suspended from a plate. Appropriate crystalline arrangements of such gyroscopes break time-reversal symmetry, opening topological phononic band gaps and supporting robust chiral edge modes [17, 119].

Unlike trivial insulators, whose electronic states can be thought of as a sum of independent local insulating states, topological insulators require the existence of delocalized states in each nontrivial band and prevent a description in terms of a basis of localized Wannier states [141–143]. It is natural, therefore, to assume that some regularity over long distances may be key to topological behavior, even if topological properties are robust to the addition of disorder. However, the extent to which spatial order needs to be built into the structure that gives rise to topological modes is unclear. We report a recipe for constructing amorphous arrangements of interacting gyroscopes—structurally more akin to a liquid than a solid—that naturally support topological phonon spectra. By simply changing the local connectivity, we can tune the chirality of edge modes to be either clockwise or counter-clockwise, or even create both clockwise and counter-clockwise edge modes in a single material. This shows that topology, a nonlocal property, can naturally arise in materials for which the only design principle is the local connectivity. Such a design principle lends itself to imperfect manufacturing and self-assembly. Although our construction arises naturally in mechanical metamaterials, we show that it extends to electronic systems in the tight binding limit.

6.2 Amorphous Voronoi Networks

Starting from an arbitrary point set, a natural way to form a network is to generate a Voronoi tessellation, either via the Wigner–Seitz construction or by connecting centroids of a triangulation [144]. Treating the edges of the cells as bonds and placing gyroscopes at the vertices leads to a network reminiscent of 'topological disorder' in electronic systems [145], shown in Fig. 6.1a. For these amorphous networks, a range of frequencies arises in which all modes are tightly localized, and this frequency region overlaps with the corresponding band gap of the honeycomb lattice. Crucially, we find that gyroscope-and-spring networks constructed in this way from arbitrary initial point sets invariably have such a mobility gap in a frequency range determined by the strength of the gravitational pinning and spring interactions.

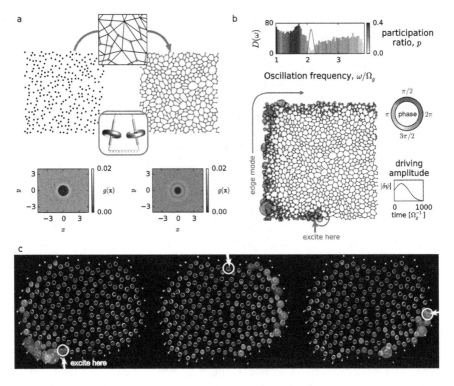

Fig. 6.1 Local structure gives rise to chiral edge modes. (**a**) Voronoization of an amorphous structure, constructed by connecting adjacent centroids of a triangulation, preserves isotropy and lack of long-range order, here with a hyperuniform point set. Two-point correlation functions $g(\mathbf{x})$ (below) are shown for a system of $N \approx 3000$ particles. (**b**) Simulations reveal chiral edge modes in topological gyroscopic networks. The localization of modes is probed by participation ratio, $p = \left(\sum_i |\psi_i|^2\right)^2 / N \sum_i |\psi_i|^4$, and the density of states is plotted as a function of normal mode oscillation frequency, in units of the gravitational precession frequency, $\Omega_g = \ell m g / I \omega_0$. The blue curve overlaying over the density of states denotes the frequency of the driving excitation in the simulation. Here, the characteristic spring frequency, $\Omega_k = k\ell^2 / I \omega_0$ is chosen such that $\Omega_g = \Omega_k$. The inset on the right shows the amplitude, $|\delta\psi|$, of the displacement for the single gyroscope which is shaken at a constant frequency. (**c**) Edge mode propagating in an amorphous experimental gyroscopic network. The motor- driven gyroscopes couple via a magnetic dipole-dipole interaction. Despite the nonlinear interaction and spinning speed disorder ($\sim 10\%$), the edge mode appears, no matter where the excitation is initialized

Our networks are reminiscent of 'topologically disordered' electronic systems [145]. In these systems, a central characteristic is that the local density of states as a function of frequency is predictive of the global density of states. Specifically, band gaps or mobility gaps are preserved [145–147]. Interestingly, we find that, even in the presence of band topology, averaging the local density of states over mesoscopic patches (~ 10 gyroscopes) reproduces the essential features of the global density of states as a function of frequency. Furthermore, we find that inserting mesoscopic patches of our structures into a variety of other dissimilar networks does not significantly disrupt the averaged local density of states of the patch.

Crucially, we find that our structures show hallmarks of non-trivial topology. When the system is cut to a finite size, modes confined to the edge populate the mobility gap, mixed in with localized states. As shown in the direct simulations of Fig. 6.1, shaking a gyroscope on the boundary results in chiral waves that bear all the hallmarks of protected edge states (robustness to disorder and absence of back-scattering).

An experimental realization can be readily constructed from gyroscopes interacting magnetically, as seen from below in Fig. 6.1c. Like in [17], these gyroscopes are constructed from 3D-printed units encasing DC motors which interact via magnetic repulsion. Probing the edge of this system immediately generates a chiral wave packet localized to the boundary, confirming that this class of topological material is physically realizable and robust (Fig. 6.1c).

This behavior begs for a topological characterization, even though it might be surprising that topology can emerge from such a local construction. The existence of chiral edge states in an energy gap is guaranteed if an invariant known as the Chern number is nonzero, and the direction of the chiral waves is given by its sign. Although the Chern number was originally defined in momentum space, several generalizations have been constructed in coordinate space in order to accommodate disorder in crystalline electronic materials [148–150]. In these methods, information about the system's vibrations above a cutoff frequency, ω_c, is carried by the projection operator, P. Each element P_{ij} measures the response of gyroscope j to excitations of gyroscope i within a prescribed range (band) of frequencies.

According to one such method, proposed in [148], a subset of the system is divided into three parts and labeled in a counterclockwise fashion (red, green, and blue regions in Fig. 6.2). These regions are then used to index components of an antisymmetric product of projection operators:

$$\nu(P) = 12\pi i \sum_{j \in A} \sum_{k \in B} \sum_{l \in C} \left(P_{jk} P_{kl} P_{lj} - P_{jl} P_{lk} P_{kj} \right). \tag{6.1}$$

The sum of such elements converges to the Chern number of the band above a chosen cutoff frequency, ω_c, when the summation region has enclosed many gyroscopes, while still being small enough to not include material close to the boundary (with respect to the localization length).

6.3 Interpretation of the Real-Space Chern Number

Equation 6.1 can be understood as a form of charge polarization in the response of an electronic material to a locally applied magnetic field. Applying a magnetic field to a small region of a material induces an electromotive force winding around the site of

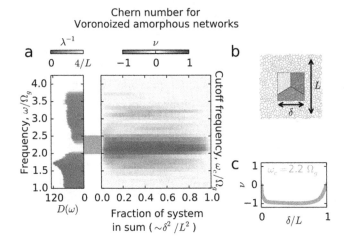

Fig. 6.2 Chern number calculations confirm topological mobility gaps. (**a**) The Chern number is computed for the band of frequencies above a cutoff frequency, ω_c, using a real-space method. Once all modes in a band that carry Hall conductance are included, the Chern number converges to an integer value. On the left is an overlaid density of states $D(\omega)$ histogram for ten realizations of Voronoized hyperuniform point sets ($\sim$2000 particles), with each mode colored by its localization length, λ. The topological mobility gaps remain in place and populated by highly localized states for all realizations. (**b–c**) The computed Chern number converges once $\sim$20–40 gyroscopes are included in the summation region (red, green, blue regions panel (**b**)), and remains at an integer value until the summation region begins to enclose the sample boundary. All networks have their precession and spring frequencies set to be equal ($\Omega_g = \Omega_k$)

application. If the material is a trivial insulator, any changes in charge density there arise from local charge re-arrangements, which result in no accumulation of charge. By contrast, a topological electronic system has a Hall conductivity determined by the Chern number. As a result, a net current will flow perpendicular to the electromotive force, inducing a net-nonzero charge concentrated at the magnetic field site, compensated by charge on the boundary. We show in Appendix D that the amount of local charging is proportional to the applied field, and the proportionality constant is the Chern number of Eq. 6.1.

Figure 6.2a shows the results of Eq. 6.1 computed for the Voronoized networks. As the cutoff frequency for the projector is varied (here it is lowered from $4\Omega_g$), the computed Chern number converges to $\nu = -1$ when all extended states in the top band lie above the cutoff frequency, confirming that the modes observed in Fig. 6.1b, c are topological in origin and predicting their direction. The Chern number remains at its value of $\nu = -1$ for a broad range of frequencies in which any existing states are localized, and thus do not contribute to the Chern number. The Chern number returns to zero once more conductance-carrying extended states are included in the calculation.

6.4   Local Geometry Controls Band Topology

Having established this connection, we now discuss how the Chern number can be controlled. In particular, we show that by considering alternative decorations of the same initial point set, it is possible to flip the chirality of the edge modes or even provide multiple gaps with differing chirality. One possible construction arises naturally from joining neighboring points in the original point set, leading to a Delaunay triangulation. Such networks show no gaps and no topology, suggesting that the local geometry dictated by Voronoization is responsible for its emergent topology. A clue can be found by noting that the Voronoized networks are locally akin to a honeycomb lattice. The honeycomb is the simplest lattice with more than one site per unit cell, a necessary condition for supporting a band gap in a lattice. Moreover, this lattice was previously found to be topological with the same Chern number [17].

Building on this insight, we introduce a second decoration, which we dub 'kagomization,' shown in Fig. 6.3a. If applied to a triangular point set, Voronoization produces a honeycomb lattice and kagomization produces a kagome lattice, the simplest lattice with three sites per unit cell, which we have found to produce $\nu = +1$ gyroscopic metamaterials. Proceeding as with the Voronoized network case (Fig. 6.3b, c), we find the presence of topologically protected modes with opposite direction and the corresponding opposite Chern number in the band structure. Other local constructions, such as the 'spindle' networks in Fig. 6.3d, e provide multiple mobility gaps, each with a different edge mode chirality, offering a transmission direction tuned by frequency.

One might think there could be a mapping from the geometry of each vertex to the chirality of the edge modes. However, taken together, our Voronoized, kagomized, and spindle networks demonstrate that simply counting nearest neighbors is not sufficient to determine the topology: a somewhat longer range description is necessary. On the other hand, we are able to change the Chern number of a structure via local decorations. To uncover the extent to which a network's topology is stored locally, consider the projection operator, P. The projector value P_{ij} measures the vibrational correlation between gyroscope j and gyroscope i when considering all modes above a cutoff frequency. By explicitly computing its magnitude in our networks, we find that the magnitude of P_{ij} falls off exponentially with distance, as shown in Fig. 6.4. Remarkably, explicitly cutting out a section of the network and embedding it in a network with a different spectrum results in only a slight change to the local projector values ($<2\%$). Since the Chern number is built from these projector elements, it then follows that the local structure of the gyroscope network, combined with some homogeneity of this local structure across the lattice, is all that is needed to determine the Chern number.

This situation is reminiscent of electronic glasses in which the local binding structure gives rise to a local 'gap.' Under weak assumptions of homogeneity, this gap can be shown to extend to the whole system [145, 146]. The case with topology is similar: the next-nearest neighbor angles in a network's cell open a local 'gap' by breaking time reversal symmetry.

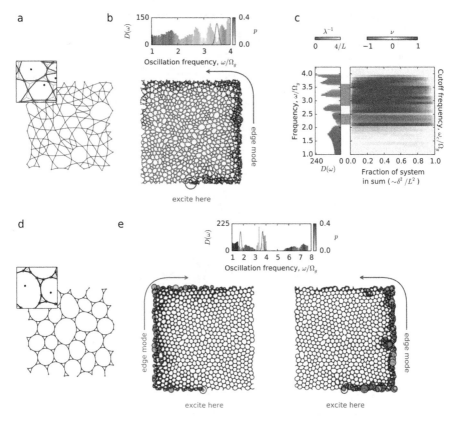

Fig. 6.3 Alternative local decorations allow control of the edge mode chirality. (**a–c**), Kagomization of an arbitrary point set yields edge modes in gyroscopic networks with the opposite chirality as Voronoized networks. (**d–e**), Another local decoration of the initial point set allows for multiple gaps with either chirality. The amorphous 'spindle' network has two gaps with chiral edge modes: blue and red curves overlaying the density of states, $D(\omega)$, mark the excitation amplitude as a function of frequency for the two cases. In panels (**b**) and (**c**), the spring frequency $\Omega_k = k\ell^2/I\omega_0$ is set equal to the gravitational precession frequency, Ω_g, while in (**e**), we chose $\Omega_k = 7\Omega_g$ to broaden the lower (clockwise) mobility gap

6.5 Spectral Flow Through Adiabatic Pumping

For amorphous networks, we make the correspondence between the bulk topological invariant and the edge states on the boundaries by considering a gyroscopic sample shaped into an annulus (c.f. [133, 151, 152]). Adiabatically tuning the interactions between pairs of gyroscopes along a radial cut (by adding a fixture to one gyro from each pair) pumps each edge mode into a neighboring mode. If we consider all states below a gap cutoff frequency, the process—which mimics the effect of threading a magnetic field through an electronic annulus' center—trades one state localized on the outer boundary for an extra state on the inner boundary of the annulus. Below, we connect this phenomenon to the real-space Chern number (Eq. 6.1).

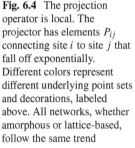

Fig. 6.4 The projection operator is local. The projector has elements P_{ij} connecting site i to site j that fall off exponentially. Different colors represent different underlying point sets and decorations, labeled above. All networks, whether amorphous or lattice-based, follow the same trend

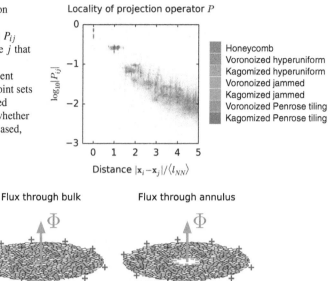

Fig. 6.5 Connecting adiabatic charge pumping to the Kitaev measure. In the bulk, the insertion of magnetic flux leads to an accumulation of charge on the inner boundary (balanced by missing charge far away). If the sample is an annulus, the flux insertion creates an extra state on the inner boundary (and removes a state on the outer boundary)

Looking ahead, we wish to establish a bulk-boundary correspondence—namely, that the number of chiral branches connecting two bands is equal to the real-space generalization of Chern number given by Eq. 6.1. We accomplish this by adiabatic pumping the spectrum of the gyroscopic networks such that an additional state is transferred to the interior boundary of an annulus. The accumulation of charge at the site of magnetic flux insertion is very similar to the occupation of an additional state localized to the inner boundary of an annulus. As shown in Fig. 6.5, threading magnetic flux in the bulk differs from threading flux through an annulus only by the absence of material in the immediate vicinity of the flux insertion site. As magnetic flux is inserted in the center of an annulus, the sample's spectrum will deform. If the Fermi level lies in a gap such that the Chern number of the lower band(s) is $\nu = +1$, threading a quantum of magnetic flux leads to one additional occupied state on the inner boundary (and one state lost on the outer boundary). This is analogous to an accumulation of charge at the insertion site (balanced by missing charge far away).

For amorphous networks, we know that the Chern sum is equal to the charge accumulated when a quantum of magnetic flux is inserted, as shown in the proof of the previous section. By introducing a hole at the site of insertion, we will now see that the real-space Chern number is also equal to the number of edge states that accumulate on the inner boundary. This argument shows the connection between the bulk topological invariant and the edge states on the boundaries by pumping edge states across the gap as an effective magnetic flux is introduced in the system [133,

151, 152]. The effective magnetic flux is supplied by a tunable alteration in the connection between a small subset of gyroscopes in the system, which introduces a phase shift in the interactions. We study the spectrum as we tune this effective phase shift.

In the electronic case, adiabatically threading a magnetic field through the center of an annulus of Chern insulator will alter the number of chiral states on the inner and outer boundaries if the Fermi level lies in the gap. Formally, this is accomplished by introducing a phase shift in the hoppings for any path enclosing the annulus' center arising from the vector potential from the threaded flux. The process will deform the system's energy spectrum such that gap states confined to one edge will rise in energy, while states confined to the opposite edge lower in energy. Once a full flux quantum is threaded, the spectrum returns to its original form, but in the process, ν chiral edge states localized to one boundary are lost to the top band while ν states are gained from the bottom band. Meanwhile, on the opposing boundary, ν chiral edge states are gained from the top and ν are lost to the bottom band (see Fig. 6.6c). This implies that on each edge, there are ν chiral channels connecting the two bands, along which states are pumped, establishing a connection between a measure of the boundary and one of the bulk [153]. There are ν states (charges) pumped to the inner boundary; we will see that this is equal to the real-space Chern number (given by Eq. 6.1) by the proof given in Appendix D.

Below, we construct an analogous argument for our amorphous systems. This is possible since the argument does not require a periodic lattice in the radial direction of the annular sample, provided the Chern number can be measured for the bulk without periodicity. By computing the spectral flow of our amorphous systems, we establish a connection between existence of our edge states and the bulk Chern number.

Consider an annular sample, such as that shown in Fig. 6.6a. The phase shift discussed above corresponds to an alteration in the interactions, such that the force of one gyro on its neighbor is altered by a rotation

$$F \sim \psi_i - \psi_j \rightarrow \psi_i - \psi_j e^{i\theta_{\text{twist}}}. \tag{6.2}$$

To give a concrete picture of how this could be built in an experiment, we envision attaching an extensible ring to a small number of gyroscopes, as illustrated in Fig. 6.6b. Since we are mimicking the effects of a gauge field, we are free to concentrate the modifications to the spring attachments along a radial cut of the annulus, shown as a blue dashed line in Fig. 6.6a, effectively supplying a 'twisted' boundary condition to the system along the cut of the annulus.

This equation of motion is then modified as follows: if the few gyroscopes just above the cut are denoted as a set, A, and their neighbors just below the cut are denoted as B, we have

$$i\dot{\psi}_i = -\Omega_g \psi_i - \frac{\Omega_k}{2} \left[\sum_{j \in \text{NN}(i)} (\chi_i - \phi_j) + e^{2i\theta_{ij}} (\bar{\chi}_i - \bar{\phi}_j) \right], \tag{6.3}$$

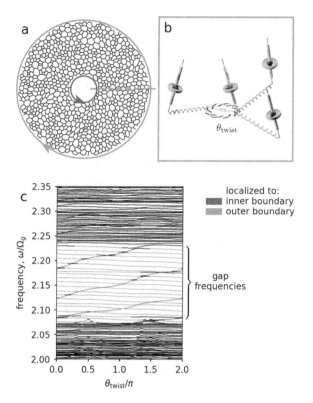

Fig. 6.6 Adiabatic pumping in an amorphous gyroscopic metamaterial. (**a**) An annulus of gyroscopes is cut along a radial line (blue dashed line). The severed connections are rejoined, but attached to an extensible ring on the gyroscopes immediately above the cut (built out of rigid rods using a Hoberman ring). The center of the ring is fixed to be directly below the gyroscope. (**b**) The point at which the new attachment connects to the gyroscopes immediately above the cut is given by the gyroscope's current displacement rotated by θ_{twist}. (**c**) The gap states in the spectrum localized to the inner boundary of the annulus rise in frequency as θ_{twist} increases, while the states on the outer edge decrease. Once the attachment point 'leads' the gyro's displacement by 2π, the spectrum has returned to its original form, but each edge state has been pumped into an adjacent state

where

$$\phi_j = \begin{cases} \psi_j e^{i\theta_{\text{twist}}} & j \in A \text{ and } i \in B \\ \psi_j & j \notin A \text{ or } i \notin B, \end{cases} \tag{6.4}$$

$$\chi_i = \begin{cases} \psi_i e^{i\theta_{\text{twist}}} & i \in A \text{ and } j \in B \\ \psi_i & i \notin A \text{ or } j \notin B, \end{cases} \tag{6.5}$$

and Ω_g and Ω_k are the gravitational precession and spring frequencies, respectively.

As the angle swept out by the attachment point is increased slowly from zero to 2π, the eigenfrequencies of the system evolve. Figure 6.6c shows the trajectories that they follow. The frequencies of states localized to the outer boundary decrease in frequency (green curves in Fig. 6.6c), while states localized to the inner boundary increasing in frequency (violet curves in Fig. 6.6c). Once the attachment point has extended a full 2π radians, the system is identical having no ring at all. Yet, if we had initialized the system in a state localized to the inner (outer) boundary, the final state would be oscillating in a normal mode with a higher (lower) frequency than the initial state. As shown in Fig. 6.6c, the nth edge state on the inner boundary evolves to the $(n + 1)$th edge state.

If we consider all states below a gap cutoff frequency, these states gain an extra state on the inner boundary and lose one state on the outer boundary. The difference between the number of final and initial states on the outer boundary, $(n - 1) - n = -1$, is precisely the Chern number of the sample's top band, since it counts the sets of states that connect the lower to the upper band.

Like the insertion of flux in the bulk, the twisted boundary condition creates the mechanical analogue of an electromotive force, directing energy in the radial direction due to the transverse (Hall) conductance, σ_{xy}. Since σ_{xy} is proportional to the Chern number, the two descriptions are equivalent. The result of the calculation is that ν states are transported to the site of flux insertion. Here, we transport ν states to the inner boundary, and the two descriptions are equal in the limit that the inner boundary is small (see Fig. 6.5).

6.6 Broken Time Reversal Symmetry

As in the lattice case, a mobility gap becomes topological due to time reversal symmetry breaking: bond angles in these networks are not multiples of $\pi/2$ (c.f. [17]). We can probe this mechanism by eliminating a gap's topology. Alternating the gravitational precession frequency, Ω_g, of neighboring gyroscopes in a network mimics the breaking of inversion symmetry on a local scale, an effect which competes against the time reversal gap opening mechanism. When the precession frequency difference between sites is large enough, this competing mechanism eliminates edge modes, triggering a transition to a $\nu = 0$ mobility gap, shown in Fig. 6.7a.

Equipped with these insights, we can easily engineer networks which are heterogeneous mixtures of multiple local configurations. Figure 6.7b–d highlight some results of combining Voronoized and kagomized networks or encapsulating one within another. Because the Voronoized and kagomized networks share a mobility gap, excitations are localized to their interface, offering a method of creating robust unidirectional waveguides, such as the sinuous waveguide shown in Fig. 6.7b. Figure 6.7c demonstrates that additional topological mobility gaps at higher frequency in the kagomized network allow bulk excitations to be confined to an encapsulated Voronoized region. Random mixtures of the two decorations,

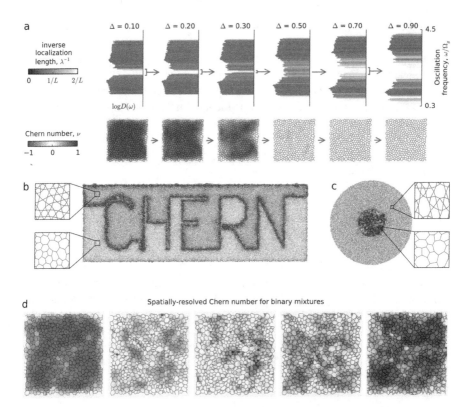

Fig. 6.7 Transition of a topological amorphous network to the trivial phase, and binary mixtures of Voronoized and kagomized networks. (**a**) Locally breaking inversion symmetry by increasing and decreasing the precession frequencies of alternating gyroscopes competes with broken time reversal symmetry, triggering a transition to the trivial phase, with no edge modes. The precession frequency splitting, Δ, is tuned so that $\Omega_g^A = \Omega_k(1 + \Delta)$ and $\Omega_g^B = \Omega_k(1 - \Delta)$. (**b**) Edge modes are localized at the interfaces between kagomized and Voronoized networks, permitting sinuous channels for the propagation of unidirectional phonons. (**c**) Excitations of a Voronoized region nested inside a kagomized network remain confined when the excitation frequency is in a mobility gap unique to the kagomized network. (**d**) When kagomized elements are randomly mixed into a Voronoized network, the sign of the local, spatially-resolved Chern calculation is determined by the local geometry, with excitations in a mobility gap biased toward the interface of the two clusters

shown in Fig. 6.7d, demonstrate heterogeneous local Chern number measurements, with mobility-gap excitations biased toward the interfaces between red and blue regions.

As our networks are structurally akin to liquids, they support topological modes in the absence of long range spatial order. The details of the underlying point set are not essential, and neither are the details of the local Voronoization or kagomization procedures. We verified this by replacing the centroidal construction [144] with a Wigner–Seitz construction. Beyond mechanical materials, we find similar results in electronic tight binding models of amorphous networks, underscoring the generality of the finding. In particular, we find similar behavior in an amorphous electronic tight binding model using the model Hamiltonian

$$H = -t_1 \sum_{\langle ij \rangle} c_i^\dagger c_j - t_2 \sum_{\langle\langle ij \rangle\rangle} e^{-i\phi_{ij}} c_i^\dagger c_j, \qquad (6.6)$$

where $\langle ij \rangle$ denotes nearest neighbors ij and $\langle\langle ij \rangle\rangle$ denotes pairs of next-nearest neighbors (NNN). The parameter t_2 tunes the strength of all NNN hoppings, and ϕ_{ij} controls the degree to which the hopping $i \to j$ breaks time reversal symmetry (by tuning the imaginary term). We find that topological edge modes arise in amorphous tight binding networks, as shown in Fig. 6.8. The main result is indifferent to the choice of NNN hopping dependence on geometry, so long as it breaks time reversal symmetry.

Fig. 6.8 Edge modes in an amorphous electronic Chern insulator. Topological edge modes arise for an electronic tight binding model of the form given by Eq. 6.6. (**a**) Topological edge modes arise in amorphous networks in which $\phi_{ij} = \pm\pi/2$ for all clockwise (counterclockwise) hoppings. Similar behavior results from choosing ϕ_{ij} to depend on the geometry of the network: in panel (**b**) $\phi = 2\theta_{nml}$, where θ_{nml} is the bond angle at vertex m connecting site n to its NNN l. Panels (**c**) and (**d**) show nearly identical spatially-resolved Chern number measurements that confirm the edge modes' topological origin

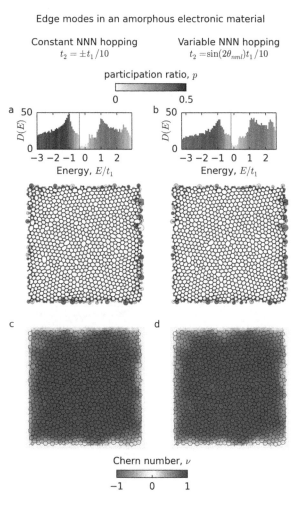

6.7   Conclusion

This study demonstrates that local interactions and local geometric arrangements are sufficient to generate chiral edge modes, promising new avenues for engineering topological mechanical metamaterials generated via imperfect self-assembly processes. Such self-assembled materials could be constructed, for instance, with micron-scale spinning magnetic particles. Since our methods bear substantial resemblance to tight-binding models, our results also find direct application not only to electronic materials, as we have demonstrated, but also to photonic topological insulators [117], acoustic resonators [129, 139], and coupled circuits [125].

Chapter 7
Conclusions and Outlook

In this work, we have demonstrated two ways in which geometry governs the mechanics of materials. Here we offer avenues for further investigation and ask questions beyond the scope of this thesis.

In Part I, we found that substrate curvature guides material failure, both in macroscopic sheets draped on curved surfaces and in nanoparticle sheets stamped to lattices of spheres. Several extensions of the work in Chap. 2 remain:

- Our perturbation theory treatment of crack paths in sheets draped on curved surfaces could be improved by including higher order terms of the Williams expansion for the stress. This could enable a more accurate perturbation theory approach, at the cost of increased computational complexity.
- The phase field model we used extended the KKL model—perhaps the simplest phase field model for fracture—to curved surfaces. Extending our phase field model to include thermodynamic consistency and modeling the process in its fully three dimensional form would increase the accuracy of the simulations, perhaps also leading to interesting physics [154].
- In ongoing work, we have begun to explore oscillatory crack instabilities in thin elastic sheets triggered by curvature.
- We also have begun exploring the ways in which curved surfaces alter crack-crack interactions.
- Our treatment throughout Chap. 2 was entirely quasistatic: the stress field is allowed to rearrange and relax at each increment of crack extension. However, under extreme circumstances, cracks in elastomers can travel at speeds comparable to the shear and longitudinal wave speeds, requiring a dynamic description [155, 156]. Are there general principles to be learned by studying the *dynamics* of fracture on curved surfaces?
- While we have studied the effect of given Gaussian curvature distributions on crack paths, a holistic approach to engineering cracks should also include a study of how to design curved surfaces in order to trigger desired crack behaviors.

© Springer Nature Switzerland AG 2020

N. Mitchell, *Geometric Control of Fracture and Topological Metamaterials*,
Springer Theses, https://doi.org/10.1007/978-3-030-36361-1_7

Inverting the problem to *design* fracture patterns—using evolutionary algorithms, for instance—remains an open challenge.

Chapter 3 presented experimental and theoretical results on stamping nanoparticle sheets to lattices of spheres. Changing the radius of curvature of the spheres enabled us to tune through two orders of magnitude in the corrugated substrate's Gaussian curvature, dialing through three regimes of resulting nanoparticle morphology. Several open questions and future goals arise for this work:

- What are the dynamics of the nanoparticle sheet's rigidity during the stamping process? As water vaporizes, leaving the dodecane-thiol matrix, how does the sheet's elastic modulus evolve, and how does its strengthening correlate with stamping progress and material failure? Does the material create dislocations while water is still present in the matrix, which would lower the energetic barrier for dislocation mobility?
- Ongoing and future work here will establish a self-consistent analytical formulation of the elastic problem with strong pinning forces adhering the elastic sheet as it stamps onto a substrate.
- A long term vision for this avenue of research would establish the strengths and weaknesses of using substrate curvature with adhesion as a tool for guiding and patterning both elastic deformation and material failure.

In Part II, we found that simple networks of interacting gyroscopes readily give rise to topological phononic band structures. In Chap. 4, we studied a real-time topological phase transition experimentally. Chapter 5 extended this analysis to many lattices, including lattices whose band structures can be tuned through phase transitions, either through bond-length-preserving deformations of the lattice or through variations of the bond strengths or pinning strengths. Beyond periodic systems, we discovered in Chap. 6 that amorphous networks—structurally akin to liquids or glasses—support topologically nontrivial band structure. These systems' band topology is measured through a real-space generalization to the Chern number. The studies of Chaps. 4, 5, and 6 prompt as many questions as they answer.

- Ongoing work hints that the universality class of the topological Anderson insulator transition is consistent with that expected for electronic insulators in symmetry class A. Are there qualitative differences in the renormalization group flow between the gyroscopic case from the two band tight-binding model?
- Additional ongoing work shows that solitons and dark solitons are possible in these systems. Can these be realized experimentally? What distinguishes solitons in a gyroscopic system from solitons of the nonlinear Schrödinger equation?
- What sets the position band gaps? In the absence of translational symmetry, how can one know *a priori* if a mobility gap will be present?
- Can $\mathbb{Z}_2$ topological insulators be built from gyroscopic components?
- Can floquet topological insulators, with time taking the place of one or more of the spatial dimensions, be constructed from gyroscopes?
- If we stack degrees of freedom, such as by building double gyroscopic pendula, what interesting behaviors can arise?

- If we link gyroscopes in 3D using stacks of coupled lattice planes, can we build Weyl semimetals?
- The dynamical laws of gyroscopic networks have symplectic structure. Looking beyond topological behavior, what consequences does this structure have for our metamaterials?

Appendix A
Creation of Surfaces of Revolution with Prescribed Gaussian Curvature

In Chap. 2, we built surfaces of prescribed Gaussian curvature. In this Appendix, we demonstrate our approach to building a surface and construct a pseudospherical patch as an example.

A.1 Governing Equations

Choosing a semi-geodesic parametrization of the surface's metric

$$g = \begin{pmatrix} 1 & 0 \\ 0 & \phi^2 \end{pmatrix},$$

(A.1)

we emanate geodesics u from a line v, as illustrated in Fig. A.1. With this choice of metric form, if we walk along a geodesic of constant v, our path length s varies with the parametrization of the curve t. For convenience, we set

$$\left(\frac{ds}{dt}\right)^2 = E\dot{u}^2 = E = 1$$

(A.2)

where we have chosen $\dot{u} \equiv \frac{du}{dt} = 1$ and $ds/dt = 1$, and where $E \equiv \mathbf{x}_u \cdot \mathbf{x}_u$. The lines of curvature of a surface of revolution are its meridians and parallels (Fig. A.1). For our specific surface, we can choose the function ϕ appearing in the metric (Eq. A.1) to be independent of the parallel, v, so that $\phi = \phi(u)$. An expression for the Gaussian curvature on the surface gives

$$G = -\frac{\phi_{uu}}{\phi}$$

(A.3)

© Springer Nature Switzerland AG 2020
N. Mitchell, *Geometric Control of Fracture and Topological Metamaterials*,
Springer Theses, https://doi.org/10.1007/978-3-030-36361-1

Fig. A.1 A surface of revolution with prescribed Gaussian curvature is parametrized by geodesic length u and v along the meridian and parallel intersecting P. Here, $\Phi(u)$ gives the radial distance along the axis of the surface of revolution, while $\Psi(u)$ determines how quickly a path marches in the $\hat{z}$ direction with respect to its path length

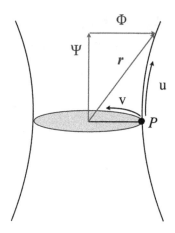

and the compatibility relations (Codazzi's equation) for the second fundamental form, b, are

$$\nabla_a b_{\beta\gamma} = \nabla_\beta b_{\alpha\gamma}. \tag{A.4}$$

Below, we solve for this surface using only Eq. A.3 and conditions on the curve v.

A.2   Equations for the Surface of a Pseudosphere

As an example, we build a surface of revolution with constant negative Gaussian curvature, known as a pseudosphere. For any semi-geodesic parametrization, we can express the curvature as Eq. A.3. Setting $G = -1$ to build a pseudosphere yields

$$\phi_{uu} = -\phi G = \phi \tag{A.5}$$

with the solution

$$\phi = Ae^u + Be^{-u}. \tag{A.6}$$

For our choice of parametrization of this surface, $\phi_u(u = 0) = 0$. That is, we choose $u = 0$ to correspond to the minimal radius from the axis of rotation to the surface, which mandates

$$\phi = Ae^u + A^{-u}. \tag{A.7}$$

At $u = 0$, $ds \equiv d\theta$ sets the scale of θ:

$$\phi(u = 0) = 1 \tag{A.8}$$

$$ds^2 = \phi^2 d\theta^2|_{u=0} \tag{A.9}$$

which results in $\phi = \cosh(u)$.

For our surface, we have a real-space surface vector (Fig. A.1)

$$\mathbf{r} = \Phi \hat{\rho}(\theta) + \Psi \hat{z} \tag{A.10}$$

$$\mathbf{r}_u = \frac{d\Phi}{du} \hat{\rho}(\theta) + \frac{d\Psi}{du} \hat{z} \tag{A.11}$$

$$\mathbf{r}_v \equiv \mathbf{r}_\theta = \Phi \hat{\theta} \tag{A.12}$$

where we are setting $\Phi \equiv \phi(u)$ and noting that $\hat{\theta} = d\hat{\rho}/dv$. In terms of the real space functions, our metric reads

$$g = \begin{pmatrix} \frac{d\Psi}{du}^2 + \frac{d\Phi}{du}^2 & 0 \\ 0 & \Phi^2 \end{pmatrix}. \tag{A.13}$$

In light of Eq. A.1, set $d\Psi/du = \sqrt{1 - (d\phi/du)^2}$ and $\Phi = \phi$ so that we recover the same form for the metric in u, v coordinates:

$$g = \begin{pmatrix} 1 & 0 \\ 0 & \phi^2 \end{pmatrix}. \tag{A.14}$$

In real space coordinates, then, we find

$$\mathbf{r} = \cosh(u)\hat{r}(v) + \Psi(u)\hat{z}, \tag{A.15}$$

$$\Psi(u) = \int_0^u \sqrt{1 - \sinh^2(u)} du' \tag{A.16}$$

where, for instance, $\hat{r}(v) = (\cos(v), \sin(v), 0)$. Note that this embedding must have a cusp somewhere, and in fact it has two. These points are where $\phi_u = 1$. The second fundamental form is no longer physical there, and one radius of curvature diverges while the other is zero.

A.3 Obtaining Geodesic Circles

Now that we have a surface, we construct a geodesic circle around the point P, as shown in Fig. A.2. A geodesic circle is a curve for which the distance along a geodesic from some point (the 'center') to any point on the curve is equal. This defines a natural boundary for a circular elastic disk stretched onto a curved surface in Chap. 2, and so we use geodesic circles as our boundaries. Here, we show our method for constructing a geodesic circle.

Fig. A.2 Coordinate system for the surface is obtained in polar coordinates (t, η) about point P. Due to the isotropic nature of the surface, the metric does not depend on η

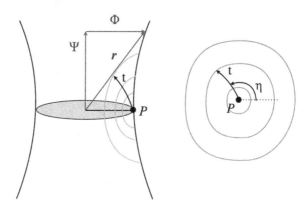

For a new coordinate system, we think intuitively of a polar coordinate system with a 'radial' coordinate t and an angular variable η. We seek a metric of the form of Eq. A.1 again in this new coordinate system, and we reuse the variable ϕ in this new context. Since $\phi(u)$ is a measurement of the 'circumference' of a geodesic circle, we see intuitively that locally near our point, we have a vanishing radius, so the geodesic circle must have the property that as $t \to 0$,

$$\phi(t \to 0) = 0. \tag{A.17}$$

We also have a smoothness criterion: as u increases, the circumference of the geodesic circle must grow as the circumference of a planar circle grows. In other words, the surface is locally flat, so for $u \to 0$,

$$\phi_t(t = 0) = 1. \tag{A.18}$$

Combining Eqs. A.17 and A.18 with Eq. A.5 (whose solution is Eq. A.6), we obtain

$$\phi(t) = \sinh t. \tag{A.19}$$

The previous calculation of the new metric form in terms of t, η gives the metric in polar coordinates around point P. We want to find geodesics on the surface emanating 'radially' ($\eta = \text{const}$) from P in terms of the coordinates u, v of our surface. To obtain an ODE giving the geodesic circle in terms of $u(t), v(t)$ for each η, we use an expression for the geodesic curvature which we derive below, along with our surface metric $\phi(u(t))$, and set it to zero along geodesics emanating from the point we have chosen. We also choose the parametrization 'speed' of the emanations $\frac{ds}{dt} = 1$, so

$$\left(\frac{ds}{dt}\right)^2 = \dot{u}^2 + \phi^2(u)\dot{v}^2 = 1 \tag{A.20}$$

Rewriting, we can express $\dot{v}$ in terms of u to look for an ODE for u,

$$\dot{v} = \sqrt{1 - \dot{u}^2}\phi^{-1}. \tag{A.21}$$

We additionally note that the Christoffel symbols, which we will need, are

$$\Gamma^1 = \begin{pmatrix} 0 & 0 \\ 0 & -\phi\phi_u \end{pmatrix} \qquad \Gamma^2 = \begin{pmatrix} 0 & \frac{-\phi_u}{\phi} \\ \frac{-\phi_u}{\phi} & 0 \end{pmatrix} \tag{A.22}$$

The definition of geodesic curvature is [157]

$$\kappa_g = \left[\Gamma^2_{11}\partial_t u^3 + (2\Gamma^2_{12} - \Gamma^1_{11})\dot{u}^2\dot{v} + (\Gamma^2_{22} - 2\Gamma^1_{12})\dot{u}\dot{v}^2 - \Gamma^1_{22}\dot{v}^3 + \dot{u}\ddot{v} - \ddot{u}\dot{v} \right] \sqrt{E\tilde{G} - F^2} \tag{A.23}$$

where $F = \mathbf{x}_u \cdot \mathbf{x}_v$ and $\tilde{G} = \mathbf{x}_v \cdot \mathbf{x}_v$. The tilde distinguishes the fundamental form $\tilde{G}$ from the Gaussian curvature, G. Here, we will build geodesics emanating from a point to parametrize our surface. By definition, the geodesic curvature κ_g vanishes along these geodesics. This allows us to set the factor in brackets in Eq. A.23 to zero. Additionally, the symmetry $\Gamma^i_{12} = \Gamma^i_{21}$ allows us to write

$$\kappa_g \propto \ddot{u}\dot{v} - \dot{u}\ddot{v} + \dot{v}(\dot{u}\ \dot{v})\Gamma^1 \begin{pmatrix} \dot{u} \\ \dot{v} \end{pmatrix} - \dot{u}(\dot{u}\ \dot{v})\Gamma^2 \begin{pmatrix} \dot{u} \\ \dot{v} \end{pmatrix}. \tag{A.24}$$

Now we use Eq. A.24 to find the ODE of the geodesic circle. Since factors of $\dot{v}\ddot{v}$ are more convenient than factors of $\ddot{v}$ alone, we multiply Eq. A.24 by $\dot{v}$ and set this to zero

$$0 = \dot{v}\kappa_g = \ddot{u}\dot{v}^2 - \dot{v}\ddot{v}\dot{u} - \dot{v}^4\phi\phi_u + 2\frac{\phi_u}{\phi}\dot{v}^2\dot{u}. \tag{A.25}$$

This gives the result

$$0 = \ddot{u} + \frac{\phi_u}{\phi}\left(-4\dot{u}^4 + 5\dot{u}^2 - 1\right). \tag{A.26}$$

Since we want to emanate the geodesics at angles η, the initial conditions are

$$\dot{u}|_{t=0} = \cos(\eta) \tag{A.27}$$

$$\dot{v}|_{t=0} = \sin(\eta) \tag{A.28}$$

for each η, and we choose point P as $u(t = 0) = 0$, $v(t = 0) = 0$, as shown in Fig. A.2. These equations are sufficient to generate patches of a surface of revolution with prescribed Gaussian curvature, bounded by geodesic circles.

Appendix B
Stretching Energy in Stamped Sheets on Spherical Surfaces

Here we address the energy scaling in a sheet stamped to the cap of a sphere. With reference to Chap. 3, we examine the particular case where the sheet is strongly pinned to the sphere as it makes contact, and we discuss how this situation differs from the analogous situation without pinning.

The integrated Gaussian curvature over a spherical cap scales as

$$2\pi \int_0^{R\theta_a} G\,r\,dr \sim R^2\theta_a^2/R^2 = R^0\theta_a^2, \tag{B.1}$$

where θ_a is the maximum polar angle included in the cap, R is the radius of the sphere, and $G = 1/R^2$ is the Gaussian curvature. A sheet conformed to such a surface experiences geometric frustration: it cannot fully relax due to the presence of the curved substrate. The geometric frustration on the spherical cap is also the source of elastic energy in an annulus of the sheet that has not yet conformed to the sphere. In particular, let us consider the portion of the sheet near θ_a which is just about to adhere to the sphere, about to become pinned in its current state of strain. The strain at θ_a scales with the integrated Gaussian curvature of the spherical cap [14, 43]:

$$\epsilon \sim \int_0^{R\theta_a} G\,r\,dr \sim R^0\theta_a^2. \tag{B.2}$$

As a result, after many annuli have adhered, each corresponding to a ever-larger θ_a, we expect $\epsilon \sim \theta^2$. Linear elasticity dictates that the stress scales similarly as well—$\sigma \sim Y\epsilon \sim Y\theta^2$, where Y is the stiffness—and thus the stretching energy density $\mathcal{E}_s = \frac{1}{2}\sigma\epsilon \sim Y\theta^4$. The total stretching energy contained in a region of nanoparticle sheet up to θ_a would then scale as $E_s = 2\pi \int_0^{R\theta_a} \mathcal{E}_s r\,dr \sim YR^2\theta_a^6$. Sequential pinning of the nanoparticle sheet ensures that this is true irrespective of the maximum angle subtended by the sheet.

© Springer Nature Switzerland AG 2020
N. Mitchell, *Geometric Control of Fracture and Topological Metamaterials*,
Springer Theses, https://doi.org/10.1007/978-3-030-36361-1

    This analysis contrasts with the expectation for an equilibrated elastic sheet without pinning. Without pinning, the energy density rearranges in such a way as to be non-monotonic in the polar angle θ on the sphere, with some sensitivity to the boundary conditions. The stress is greatest on the apex of a sphere without pinning, in stark contrast to the case with sequential pinning, for which the stress vanishes at the cap. Solving for the unpinned situation exactly results in an integrated energy density which is roughly quadratic in polar angle: $E = 2\pi \int_0^r \mathcal{E} r \, dr \sim Y R^2 \theta^2$, with corrections of order $\mathcal{O}(R^2 \theta^4)$. This difference highlights the distinct character of sequential adhesion to a substrate seen in the nanoparticle system.

Appendix C
Symplectic Structure of Gyroscopic Motion

A symplectic structure underlies the equations of motion for coupled gyroscopes suspended from fixed pivot points. When the spinning speed is large, the variations in kinetic energy become negligible as the amplitudes of nutation vanish. Therefore, the energy can be entirely captured by the potential energy stored in the interactions (springs, for example) between gyroscopes. Consider a gyroscope with hanging orientation vectors $\ell\hat{n} = (x, y)$ subjected to forces acting at $\ell\hat{n}$. Its motion is governed by

$$I\omega_0\dot{x}/\ell = -\ell F_y = \ell\frac{\partial U}{\partial y_i} \tag{C.1}$$

$$I\omega_0\dot{y}/\ell = \ell F_x = -\ell\frac{\partial U}{\partial x_i}, \tag{C.2}$$

where I is the principal moment of inertia, ω_0 is the spinning speed of the gyroscope, U is the potential energy stored in the springs connecting the gyroscope to its neighbors, $\vec{F} = -\vec{\nabla}U$ is the net force. Now we define

$$q = \sqrt{I\omega_i}x/\ell \tag{C.3}$$

$$p = \sqrt{I\omega_i}y/\ell, \tag{C.4}$$

so that Eqs. C.1 and C.2 are the same as Hamilton's equations:

$$\dot{q} = \frac{\partial U}{\partial p} \qquad \dot{p} = -\frac{\partial U}{\partial q}. \tag{C.5}$$

Here we show that this structure has consequences for computing band topology in gyroscopic lattices, which is encoded in the Chern number. To calculate the Chern number for gyroscopes on a lattice, we first find the band-structure using

© Springer Nature Switzerland AG 2020
N. Mitchell, *Geometric Control of Fracture and Topological Metamaterials*,
Springer Theses, https://doi.org/10.1007/978-3-030-36361-1

the linearized equation of motion. For each site in a unit cell, we assume a solution which is composed of clockwise and counter-clockwise propagating modes:

$$\psi_p = \psi_p^R e^{i(\vec{k}\cdot\vec{x}-\omega t)} + \tilde{\psi}_p^L e^{-i(\vec{k}\cdot\vec{x}-\omega t)}, \tag{C.6}$$

where p indexes the n sites in a unit cell. The resulting equations can be expressed as the $2n \times 2n$ dynamical matrix which is a function of the wave vector, $\vec{k}$. This dynamical matrix resembles hopping model matrices seen for lattice calculations in quantum mechanics.

Diagonalizing the dynamical matrix yields $2n$ frequencies of the dispersion bands at each value of $\vec{k}$. The eigenvalues come in positive/negative pairs, but each pair represents the same oscillation in real space. Because of this redundancy, we discuss only the positive eigenvalues for each system. At each value of $\vec{k}$, each mode has a corresponding eigenvector,

$$|u_j(\vec{k})\rangle = \left(\psi_1^R, ..., \psi_N^R, \psi_1^L, ..., \psi_N^L \right) \tag{C.7}$$

characterizing the amplitudes and phases of the N gyroscopes' collective motion. The symplectic symmetry of the dynamical matrix enables the eigenstates to be orthogonalized such that

$$\langle u_i | u_j \rangle = \sum_\alpha \overline{\psi_{i,\alpha}^R} \psi_{j,\alpha}^R - \overline{\psi_{i,\alpha}^L} \psi_{j,\alpha}^L \tag{C.8}$$

$$= \delta_{ij} \operatorname{sign}(\omega_j) \tag{C.9}$$

where α runs over each gyroscope and ω_j is the oscillation frequency of $|u_j\rangle$.

The Chern number of each band is given by an integral of the Berry curvature, $\mathcal{F}(\vec{k})$:

$$C_j = \frac{1}{2\pi} \int d^2k \, \mathcal{F}_j(\vec{k})$$

$$= \frac{i}{2\pi} \oint A_j(\vec{k}) \cdot dk, \tag{C.10}$$

where $A_j(k) = i \langle u_j | \nabla_k u_j \rangle$. In this work, Chern numbers are calculated numerically using the phase-invariant formula [132]

$$C_j dx \wedge dy = \frac{i}{2\pi} \int d^2k \, \operatorname{Tr}[dP_j \wedge P_j dP_j], \tag{C.11}$$

where C_j is the Chern number of the jth band, $\wedge$ is the wedge product, and P_j is the projection matrix defined for our system as

$$P_j = |\psi_\alpha\rangle \langle \psi_\alpha | Q \operatorname{sign}(\omega_j), \tag{C.12}$$

where

$$Q = \begin{pmatrix} \mathbb{I}_n & 0 \\ 0 & -\mathbb{I}_n \end{pmatrix}. \tag{C.13}$$

The factors of $Q \operatorname{sign}(\omega_j)$ arise from the symplectic structure of our equations of motion. Note that in previous studies of gyroscopic lattices, the projector was defined using a simple outer product of bands which were not orthonormal. Although this does not affect any of the computed Chern numbers of the bands, it leads to a non-physical distribution of Berry curvature in those bands. By using the symplectic formulation, the correct Berry curvature distributions are readily obtained.

Appendix D
Interpretation of Real-Space Chern Number

The Chern number formula of Eq. 6.1 can be intuitively understood by mapping our gyroscopic metamaterial to a model of electrons tunneling between sites in a 2D material [17]. In this interpretation, probing the Chern number by summing over all states below a selected frequency corresponds to measuring the Chern number of an electronic system with all electronic states filled below the selected energy. In the electronic system, the Chern number is proportional to the Hall conductivity, and the projection operator element, P_{ij}, is

$$P_{ij} = \sum_{n \in \text{band}} \psi_n(\mathbf{x}_i)\psi_n^*(\mathbf{x}_j), \tag{D.1}$$

where ψ_n is the wavefunction with energy E_n and the sum is over all the states with energies below the Fermi level. $|P_{ij}|^2$ is then the correlation in electron density at the two points $\mathbf{x}_i$ and $\mathbf{x}_j$ when all the states with energies below the Fermi energy are occupied.

Two salient aspects to the integer quantum Hall effect are the current induced as a response to an electric field,

$$J = \nu \frac{e^2}{h} \mathbf{z} \times \mathbf{E}, \tag{D.2}$$

and, secondly, the response to a magnetic field applied at a point in the sample. The latter takes the form of charge accumulation at the site of application according to

$$\rho = \rho_0 + \nu \frac{e^2}{h} B_z. \tag{D.3}$$

In the above equations, $\mathbf{z}$ is the normal vector to the plane, B_z is the magnetic field in that direction, ρ_0 is the charge density before the magnetic field is applied, and ν is an integer. If the system is periodic, ν is the Chern number of the occupied bands.

© Springer Nature Switzerland AG 2020
N. Mitchell, *Geometric Control of Fracture and Topological Metamaterials*,
Springer Theses, https://doi.org/10.1007/978-3-030-36361-1

It can be seen that Eqs. D.2 and D.3 are consistent by noting that the continuity equation follows from Faraday's law.

Equation D.3 can be understood intuitively as follows. Turning on a field B_z in a small region induces an electromotive force that pulls charge towards the region according to the Hall effect. A local field applied to an insulator cannot cause charge to build up in a given area of the material. This can be understood by appealing to perturbation theory: a local field merely mixes the initial state with other states that have the same number of particles, and thus cannot change the charge density. However, our point-like magnetic field is not a local field since the vector potential of this field decays to infinity slowly, as $1/r$. Therefore, turning on a magnetic field in a small region can excite dipole pairs all the way out to infinity, creating a net charge in the small region where the magnetic field is nonzero. If the sample is of finite size, the compensating, opposite charge is confined to the boundary of the sample.

In this section, we first show how the projection operator responds to a general perturbation of the Hamiltonian. We then show that the Chern number as expressed in Eq. 6.1 is exactly the proportionality constant between the amount of charge that accumulates at the site of a point-like magnetic field and the strength of that magnetic field.

Start first with a generic perturbation, $V(r)$, which could take the form of an induced potential energy or the contribution to the Hamiltonian due to a magnetic field. If the unperturbed wavefunctions are $\psi_n(r)$, then the perturbed wavefunctions, $\chi_n(r)$, are given by

$$\chi_n(r) \approx \psi_n(r) + \sum_m \frac{\langle m|V|n \rangle}{E_n - E_m + i\eta} \psi_m(r), \tag{D.4}$$

where $|n\rangle = \psi_n(r)$ is the state with energy E_n and η is an infinitesimal real number.

Consider now a network of sites x_i on which electrons may hop, so that we replace r by the label of a site i or j. The projection onto a band of energies is then

$$P_{ij} \equiv \sum_{n \in \text{band}} \chi_n(i)\chi_n^*(j), \tag{D.5}$$

so the change in the projection is

$$\Delta P_{ij} = \sum_{\substack{\text{all } m, \\ n \in \text{band}}} \frac{\langle m|V|n \rangle}{E_n - E_m + i\eta} \psi_m(i)\psi_n^*(j) + c.c., \tag{D.6}$$

where c.c. signifies the complex conjugate. A singularity in the denominator from terms for which $E_m = E_n$ would lead to a long-range sensitivity in this system. However, the terms in which E_m is in the occupied band cancel out among one another so there is no longer a singularity:

$$\Delta P_{ij} = \sum_{\substack{m \notin \text{band}, \\ n \in \text{band}}} \frac{\langle m|V|n \rangle}{E_n - E_m} \psi_m(i) \psi_n^*(j) + c.c. \tag{D.7}$$

We have dropped the $i\eta$ term by assuming that our Fermi level is in a band gap: the two nearest energies (maximum m in a band, and minimum n above it) are sufficiently far apart. The cancellation of terms in which m is in the occupied band can be seen to result from the Pauli exclusion principle: a perturbation cannot cause transitions into a state that is occupied already, so there are only contributions from terms in which E_m is in the empty bands.

Now let us see that there are only short-range effects of a perturbation when there is no singularity. At first, Eq. D.7 may appear to have long-range contributions, at least in an ordered system, because then the wavefunctions are delocalized (they extend to infinity). However, summing over all the energies in the band, Eq. D.7 becomes a Fourier transform of a product of wavefunctions, and the Fourier transform of a smooth function decays exponentially with distance.

We are interested in how the density of the system changes. The charge at site i is simply the elementary charge, e, times the sum of the probabilities for all the occupied states to be at that site, which is just $e P_{ii}$.

Specializing Eq. D.7 to the case where the system is described by a hopping model lets us write $H_{ij} = -t_{ij}$, which is the amplitude for hopping from one site to another. This case is particularly similar to our discrete gyroscopic metamaterials. A magnetic field is described by adding phase factors, $H_{ij}' = -t_{ij}e^{i\phi_{ij}}$, where the phases are chosen so that for a triangle made up of three sites, the magnetic flux through the triangle is the sum of the three phases (up to a fixed proportionality constant). If the phases are small, the magnetic field perturbation's matrix elements are $V_{ij} \approx -it_{ij}\phi_{ij}$. Substituting this into the expression for ΔP_{ii} gives an expression for the charge density. We would like to see that this expression is quantized. Somehow, the dependence on t and E must cancel out and give an integer. In fact, the t matrix can be written in terms of the energies by inserting a complete set of states:

$$t_{ij} = -\langle i|H|j \rangle = -\sum E_n \psi_n(i) \psi_n^*(j), \tag{D.8}$$

where here $|i\rangle$ denotes the wavefunction at site i. Denoting the number of charges at site i as $n(i) = P_{ii}$, this leads to

$$e \Delta n(i) = 2e \, \text{Im} \left\{ \sum_{\substack{j,k, \\ e \notin \text{band}, \\ e_1 \notin \text{band}, \\ b \in \text{band}}} \frac{\phi_{jk} E_{e_1} \psi_{e_1}^{k*} \psi_{e_1}^{j} \psi_e^{j*} \psi_e^{i} \psi_b^{i*} \psi_b^{k}}{E_e - E_b} \right.$$

$$\left. + \sum_{\substack{j,k, \\ e \notin \text{band}, \\ b \in \text{band}, \\ b_1 \in \text{band}}} \frac{\phi_{jk} E_{b_1} \psi_{b_1}^{k*} \psi_{e_1}^{j} \psi_{e}^{j*} \psi_{e}^{i} \psi_{b}^{i*} \psi_{b}^{k}}{E_e - E_b} \right\}, \qquad (\text{D.9})$$

where the b's are summed over states in the band of energies, and the e's are summed over the "excited" states outside the band. In the above expression, we have denoted $\psi_j(i)$ as ψ_j^i to streamline the notation.

Let us define $\theta_{ijk} = \phi_{ij} + \phi_{jk} + \phi_{ki}$ which is the magnetic flux through the triangle ijk, times $e/\hbar$. In the above sum, we can write $\phi_{jk} = \theta_{ijk} - \phi_{ij} - \phi_{ki}$. Then, using the fact that $\phi_{ij} = -\phi_{ji}$, all the terms depending on ϕ's cancel out. Physically, this must be so because the charge response should depend only on the physical magnetic field and not the vector potential. Proper substitution and cancellations result in an expression which is the same as Eq. D.9 but with $\phi_{jk} \rightarrow \theta_{ijk}$. The change in charge at site i decays if the magnetic field is concentrated far from i: the only triangles passing through i that enclose the magnetic field would have long sides, with corresponding projector components which are exponentially small.

Now, the response cannot depend on the dispersion of the states, only on which states are included in the band and which are not. A change in the dispersion of the bands causes a local effect (unlike the non-local effect of a magnetic field due to its vector potential), so the net charge cannot change. This observation allows us to assume that $E_b = -1$ for every state in the band and $E_e = 1$ for every state that is not in the band. For this particular dispersion, we get a simple response in terms of the projection operators:

$$e\Delta n(i) = 2e\text{Im} \sum_{jk} P_{ij} P_{jk} P_{ki} \theta_{ijk}. \qquad (\text{D.10})$$

Equation D.10 highlights the locality of the response because the amplitude of the projectors decay exponentially with distance.

The portion of Eq. D.10 which is of interest is the ratio of the net charge, $Q = e\Delta n$, to the net magnetic flux. Consider now the special case where the magnetic field is all passing through one point, so that $\theta_{ijk} = \pm\Phi_B(e/\hbar)$ if the triangle ijk encloses the magnetic flux and zero otherwise. The sign depends on whether the triangle is clockwise or counterclockwise. With a point-like magnetic field, the ratio of the net charge to the magnetic flux is

$$\frac{Q}{\Phi_B} = \frac{8\pi e^2}{h} \text{Im} \sum_{\text{triangles } ijk} P_{ij} P_{jk} P_{ki}, \qquad (\text{D.11})$$

where the sum is over only the triangles that surround the flux line in a counterclockwise way. (Since the clockwise paths are composed of the same points in the opposite order, the clockwise and counterclockwise contributions are equal.)

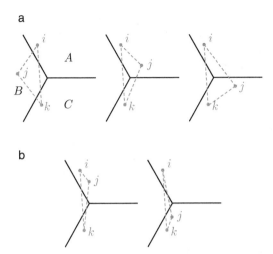

Fig. D.1 Contributions to the real-space Chern number which differ in geometry between Eqs. 6.1 and D.11. (**a**) Equation 6.1 includes terms which do not enclose the vertex shown on the left, while Eq. D.11 includes terms in which two of the vertices i, j, k lie in the same region. We find that these terms balance, so that the expressions are equivalent. (**b**) Adding terms corresponding to two additional configurations aids in showing that Eqs. 6.1 and D.11 are equivalent. In these terms, two vertices lie a single region and the triangle does not enclose the vertex

Finally, we can see that Eq. D.11 is equivalent to Eq. 6.1 after appropriate combinatorics. Note that

$$\sum_{i \in A, j \in B, k \in C} \left(P_{ij} P_{jk} P_{kl} - P_{lk} P_{kj} P_{ji} \right) = \frac{1}{3} \sum_{\substack{i,j,k \text{ from} \\ \text{distinct regions}}} P_{ij} P_{jk} P_{kl} (-1)^a,$$

$$(D.12)$$

with $a = 0$ for triangles such that the vertices i, j, k are in A, B, C or a cyclic permutation, and $a = 1$ for triangles such that i, j, k are in B, A, C or a cyclic permutation. We then see that Eqs D.11 and 6.1 differ only by certain terms shown in Fig. D.1a. These are terms in which the triangle ijk does not enclose the flux line and triangles in which two of the points i, j, and k are in the same region.

In all cases of these differing terms, one of the sides of the triangle ijk must cross two boundaries. Grouping terms in which this particular side is the same, we find the difference between expressions is a sum of terms $P_{ij} P_{jk} P_{ki} - P_{ij} P_{jk} P_{ki}$ for triangles shown in Fig. D.1a, plus cyclic permutations of regions A, B, and C and arbitrary permutations of i, j, k. If we add terms corresponding to configurations in Fig. D.1b, the sum of all configurations shown in Fig. D.1 is zero since

$$12\pi i \sum_j P_{ij} P_{jk} P_{ki} + c.c. = 12\pi i P_{ik} P_{ki} + c.c. = 0, \qquad (D.13)$$

using the fact that $P^2 = P$ and that $P_{ik} P_{ki} = |P_{ik}|^2$ is real. The added terms themselves sum to zero, implying that Eqs. 6.1 and D.11 are equivalent. The added terms vanish because these triangles each have two sides crossing through region B. When summed over all three variables i, j, k, their contribution can be written as

$$12\pi i \left(\sum_{\substack{i,j \in A, k \in C, \\ \text{s.t. } \overline{jk} \text{ and } \overline{ik} \\ \text{cross } B}} P_{ij} P_{jk} P_{ki} + \sum_{\substack{i \in A, j, k \in C, \\ \text{s.t. } \overline{ij} \text{ and } \overline{ik} \\ \text{cross } B}} P_{ij} P_{jk} P_{ki} \right)$$

$$+ c.c. + \text{permutations of } A, B, C. \qquad \text{(D.14)}$$

Since i, j can be exchanged in the first sum and j, k can be exchanged in the second sum, both are purely real, and the additional contribution vanishes. Thus, Eqs. 6.1 and D.11 are equivalent. In conclusion, we have shown the Chern number expression of Eq. 6.1 characterizes the accumulated charge at the location of an applied point-like magnetic field.

References

1. J.L.R. D'Alembert, J.S. Calero (eds.), *A New Theory of the Resistance of Fluids* (Springer, Berlin, 1752)
2. M.J. Bowick, L. Giomi, Two-dimensional matter: order, curvature and defects. Adv. Phys. **58**(5), 449–563 (2009)
3. G.A. DeVries, M. Brunnbauer, Y. Hu, A.M. Jackson, B. Long, B.T. Neltner, O. Uzun, B.H. Wunsch, F. Stellacci, Divalent metal nanoparticles. Science **315**(5810), 358–361 (2007)
4. V.K. Gupta, N.L. Abbott, Design of surfaces for patterned alignment of liquid crystals on planar and curved substrates. Science **276**(5318), 1533–1536 (1997)
5. F.C. Keber, E. Loiseau, T. Sanchez, S.J. DeCamp, L. Giomi, M.J. Bowick, M.C. Marchetti, Z. Dogic, A.R. Bausch, Topology and dynamics of active nematic vesicles. Science **345**(6201), 1135–1139 (2014)
6. S. Shankar, M.J. Bowick, M.C. Marchetti. Topological sound and flocking on curved surfaces. Phys. Rev. X **7**(3), 031039 (2017)
7. E. Winfree, F. Liu, L.A. Wenzler, N.C. Seeman, Design and self-assembly of two-dimensional DNA crystals. Nature **394**(6693), 539–544 (1998)
8. N.C. Seeman, DNA in a material world. Nature **421**(6921), 427–431 (2003)
9. A.R. Bausch, M.J. Bowick, A. Cacciuto, A.D. Dinsmore, M.F. Hsu, D.R. Nelson, M.G. Nikolaides, A. Travesset, D.A. Weitz, Grain boundary scars and spherical crystallography. Science **299**(5613), 1716–1718 (2003)
10. W.T.M. Irvine, V. Vitelli, P.M. Chaikin, Pleats in crystals on curved surfaces. Nature **468**(7326), 947–951 (2010)
11. I.R. Bruss, S.C. Glotzer, Curvature-induced microswarming. Soft Matter **13**(30), 5117–5121 (2017)
12. G. Meng, J. Paulose, D.R. Nelson, V.N. Manoharan, Elastic instability of a crystal growing on a curved surface. Science **343**(6171), 634–637 (2014)
13. R.E. Guerra, C.P. Kelleher, A.D. Hollingsworth, P.M. Chaikin, Freezing on a sphere. Nature **554**(7692), 346–350 (2018)
14. V. Vitelli, J.B. Lucks, D.R. Nelson, Crystallography on curved surfaces. Proc. Nat. Acad. Sci. **103**(33), 12323–12328 (2006)
15. W.E. Baker, Axisymmetric modes of vibration of thin spherical shell. J. Acoust. Soc. Am. **33**(12), 1749–1758 (1961)
16. R. Süsstrunk, S.D. Huber, Observation of phononic helical edge states in a mechanical topological insulator. Science **349**(6243), 47–50 (2015)
17. L.M. Nash, D. Kleckner, A. Read, V. Vitelli, A.M. Turner, W.T.M. Irvine, Topological mechanics of gyroscopic metamaterials. Proc. Nat. Acad. Sci. **112**(47), 14495–14500 (2015)

© Springer Nature Switzerland AG 2020
N. Mitchell, *Geometric Control of Fracture and Topological Metamaterials*,
Springer Theses, https://doi.org/10.1007/978-3-030-36361-1

18. S.D. Huber, Topological mechanics. Nat. Phys. **12**(7), 621–623 (2016)

19. M. Fruchart, D. Carpentier, An introduction to topological insulators. C. R. Phys. **14**(9), 779–815 (2013)

20. Google. Missouri Botanical Garden, Shaw Blvd, St. Louis, MO, (2018)

21. J.-Y. Sgro, Human Papillomavirus 16 L1 (T=7d) capsid Model, PBD code 1l0t (2009)

22. Y. Modis, B.L. Trus, S.C. Harrison, Atomic model of the papillomavirus capsid. EMBO J. **21**(18), 4754–4762 (2002)

23. SYBYL, Tripos International, Tripos.com

24. M. Tarini, P. Cignoni, C. Montani, Ambient occlusion and edge cueing for enhancing real time molecular visualization. IEEE Trans. Vis. Comput. Graph. **12**(5), 1237–1244 (2006)

25. J. Bardeen, L.N. Cooper, J.R. Schrieffer, Microscopic theory of superconductivity. Phys. Rev. **106**(1), 162–164 (1957)

26. M. Wyart, S.R. Nagel, T.A. Witten, Geometric origin of excess low-frequency vibrational modes in weakly connected amorphous solids. Europhys. Lett. **72**(3), 486–492 (2005)

27. N. Xu, V. Vitelli, M. Wyart, A.J. Liu, S.R. Nagel, Energy transport in Jammed sphere packings. Phys. Rev. Lett. **102**(3), 038001 (2009)

28. D.R. Smith, J.B. Pendry, M.C.K. Wiltshire, Metamaterials and negative refractive index. Science **305**(5685), 788–792 (2004)

29. S.A. Cummer, J. Christensen, A. Alù, Controlling sound with acoustic metamaterials. Nat. Rev. Mater. **1**(3), 16001 (2016)

30. R. Lakes, Foam structures with a negative poisson's ratio. Science **235**(4792), 1038–1040 (1987)

31. J.W. Rocks, N. Pashine, I. Bischofberger, C.P. Goodrich, A.J. Liu, S.R. Nagel, Designing allostery-inspired response in mechanical networks. Proc. Natl. Acad. Sci. **114**, 201612139 (2017)

32. V.G. Veselago, The electrodynamics of substances with simultaneously negative values of ϵ and μ. Sov. Phys. Uspekhi **10**(4), 509 (1968)

33. J.B. Pendry, A.J. Holden, D.J. Robbins, W.J. Stewart, Magnetism from conductors and enhanced nonlinear phenomena. IEEE Trans. Microw. Theory Tech. **47**(11), 2075–2084 (1999)

34. D.R. Smith, W.J. Padilla, D.C. Vier, S.C. Nemat-Nasser, S. Schultz, Composite medium with simultaneously negative permeability and permittivity. Phys. Rev. Lett. **84**(18), 4184–4187 (2000)

35. R.A. Shelby, D.R. Smith, S. Schultz, Experimental verification of a negative index of refraction. Science **292**(5514), 77–79 (2001)

36. C.L. Kane, T.C. Lubensky, Topological boundary modes in isostatic lattices. Nat. Phys. **10**(1), 39–45 (2013)

37. C. He, X. Ni, H. Ge, X.-C. Sun, Y.-B. Chen, M.-H. Lu, X.-P. Liu, Y.-F. Chen, Acoustic topological insulator and robust one-way sound transport. Nat. Phys. **12**(12), 1124–1129 (2016)

38. M. Brun, I.S. Jones, A.B. Movchan, Vortex-type elastic structured media and dynamic shielding. Proc. R. Soc. A **468**, rspa20120165 (2012)

39. R. Fleury, D.L. Sounas, C.F. Sieck, M.R. Haberman, A. Alù, Sound Isolation and giant linear nonreciprocity in a compact acoustic circulator. Science **343**(6170), 516–519 (2014)

40. C. Coulais, D. Sounas, A. Alù, Static non-reciprocity in mechanical metamaterials. Nature **542**(7642), 461–464 (2017)

41. A. Souslov, K. Dasbiswas, M. Fruchart, S. Vaikuntanathan, V. Vitelli, Topological waves in fluids with odd viscosity. Phys. Rev. Lett. **112**(12), (2019)

42. D.J. Thouless, M. Kohmoto, M.P. Nightingale, M. den Nijs, Quantized Hall conductance in a two-dimensional periodic potential. Phys. Rev. Lett. **49**(6), 405–408 (1982)

43. N.P. Mitchell, V. Koning, V. Vitelli, W.T.M. Irvine, Fracture in sheets draped on curved surfaces. Nat. Mater. **16**(1), 89–93 (2017)

44. N.P. Mitchell, R. Carey, J. Hannah, Y. Wang, M.C. Ruiz, S. McBride, X.-M. Lin, H. Jaeger, Conforming nanoparticle sheets to surfaces with Gaussian curvature. Soft Matter. **14**, 9107–9117 (2018)

45. N.P. Mitchell, L.M. Nash, W.T.M. Irvine, Realization of a topological phase transition in a gyroscopic lattice. Phys. Rev. B **97**(10), 100302 (2018)

46. N.P. Mitchell, L.M. Nash, W.T.M. Irvine, Tunable band topology in gyroscopic lattices. Phys. Rev. B **98**(17), 174301 (2018)

47. N.P. Mitchell, L.M. Nash, D. Hexner, A.M. Turner, W.T.M. Irvine, Amorphous topological insulators constructed from random point sets. Nat. Phys. **14**(4), 380–385 (2018)

48. G.M. Grason, B. Davidovitch, Universal collapse of stress and wrinkle-to-scar transition in spherically confined crystalline sheets. Proc. Nation. Acad. Sci. **110**(32), 12893–12898 (2013)

49. D.P. Holmes, A.J. Crosby, Draping films: a Wrinkle to fold transition. Phys. Rev. Lett. **105**(3), 038303 (2010)

50. J. Hure, B. Roman, J. Bico, Wrapping an adhesive sphere with an elastic sheet. Phys. Rev. Lett. **106**(17), 174301 (2011)

51. L.I. Slepyan, Cracks in a bending plate, in *Models and Phenomena in Fracture Mechanics*. Foundations of Engineering Mechanics (Springer, Berlin, 2002), pp. 359–388

52. E.S. Folias, The stresses in a cracked spherical shell. J. Mathe. Phys. **44**(1), 164–176 (1965)

53. F. Amiri, D. Millán, Y. Shen, T. Rabczuk, M. Arroyo, Phase-field modeling of fracture in linear thin shells. Theor. Appl. Fract. Mec. **69**, 102–109 (2014)

54. S.M. Rupich, F.C. Castro, W.T.M. Irvine, D.V. Talapin, Soft epitaxy of nanocrystal superlattices. Nat. Commun. **5**, 5045 (2014)

55. M.B. Dusseault, V. Maury, F. Sanfilippo, F.J. Santarelli, *Drilling Around Salt: Risks, Stresses, and Uncertainties* (American Rock Mechanics Association, New York, 2004)

56. A.A. Griffith, The Phenomena of rupture and flow in solids. Philos. Trans. Royal Soc. London A: Mathe. Phys. Eng. Sci. **221**(582–593), 163–198 (1921)

57. R.S. Rivlin, A.G. Thomas, Rupture of rubber. I. characteristic energy for tearing. J. Poly. Sci. **10**(3), 291–318 (1953)

58. L.B. Freund, *Dynamic Fracture Mechanics* (Cambridge University Press, Cambridge, 1990)

59. C.-Y. Hui, A.T. Zehnder, Y.K. Potdar, Williams meets von Karman: mode coupling and nonlinearity in the fracture of thin plates. Inter. J. Fract. **93**(1–4), 409–429 (1998)

60. V. Vitelli, A.M. Turner, Anomalous coupling between topological defects and curvature. Phys. Rev. Lett. **93**(21), 215301 (2004)

61. H.M. Westergaard, Bearing pressures and cracks. J. Appl. Mech. Trans. ASME **6**, A49–A53 (1939)

62. B. Cotterell, J.R. Rice, Slightly curved or kinked cracks. Inter. J. Fract. **16**(2), 155–169 (1980)

63. N.I. Muskhelishvili, *Some Basic Problems of the Mathematical Theory of Elasticity* (Springer, Berlin, 1977)

64. R. Ghelichi, K. Kamrin, Modeling growth paths of interacting crack pairs in elastic media. Soft Matt. **11**(40), 7995–8012 (2015)

65. T. Fett, Stress intensity factors and T-stress for internally cracked circular disks under various boundary conditions. Eng. Fract. Mech. **68**(9), 1119–1136 (2001)

66. A. Karma, D.A. Kessler, H. Levine, Phase-field model of mode III dynamic fracture. Phys. Rev. Lett. **87**(4), 045501 (2001)

67. R. Spatschek, E. Brener, A. Karma, Phase field modeling of crack propagation. Philos. Maga. **91**(1), 75–95 (2011)

68. D.R. Nelson, L. Peliti, Fluctuations in membranes with crystalline and hexatic order. J. de Phys. **48**(7), 1085–1092 (1987)

69. H. Henry, H. Levine, Dynamic instabilities of fracture under biaxial strain using a phase field model. Phys. Rev. Lett. **93**(10), 105504 (2004)

70. V. Hakim, A. Karma, Laws of crack motion and phase-field models of fracture. J. Mech. Phys. Solids **57**(2), 342–368 (2009)

71. H. Henry, Study of the branching instability using a phase field model of inplane crack propagation. Europhys. Lett. **83**(1), 16004 (2008)

72. L.D. Landau, E.M. Lifshitz, Chapter II-the equilibrium of rods and plates, in *Theory of Elasticity*, 3rd edn. (Butterworth-Heinemann, Oxford, 1986), pp. 38–86

73. A. Logg, G.N. Wells, DOLFIN: automated finite element computing. ACM Trans. Math. Softw. **37**(2), 20:1–20:28 (2010)

74. B.A. Cheeseman, M.H. Santare, The interaction of a curved crack with a circular elastic inclusion. Inter. J. Fract. **103**(3), 259–277 (2000)

75. J.M. Yuk, J. Park, P. Ercius, K. Kim, D.J. Hellebusch, M.F. Crommie, J.Y. Lee, A. Zettl, A.P. Alivisatos, High-resolution EM of colloidal nanocrystal growth using graphene liquid cells. Science **336**(6077), 61–64 (2012)

76. N.J. Price, J.W. Cosgrove, *Analysis of Geological Structures* (Cambridge University Press, Cambridge, 1990)

77. J.A. Rogers, T. Someya, Y. Huang, Materials and mechanics for stretchable electronics. Science **327**(5973), 1603–1607 (2010)

78. A. Yuse, M. Sano, Transition between crack patterns in quenched glass plates. Nature **362**(6418), 329–331 (1993)

79. E. Sharon, E. Efrati, The mechanics of non-Euclidean plates. Soft Matt. **6**(22), 5693–5704 (2010)

80. J.D. Paulsen, V. Démery, C.D. Santangelo, T.P. Russell, B. Davidovitch, N. Menon, Optimal wrapping of liquid droplets with ultrathin sheets. Nat. Mater. **14**(12), 1206–1209 (2015)

81. Z. Yao, M. Bowick, X. Ma, R. Sknepnek, Planar sheets meet negative-curvature liquid interfaces. Europhys. Lett. **101**(4), 44007 (2013)

82. H. King, R.D. Schroll, B. Davidovitch, N. Menon, Elastic sheet on a liquid drop reveals wrinkling and crumpling as distinct symmetry-breaking instabilities. Proc. Natl. Acad. Sci. **109**(25), 9716–9720 (2012)

83. J. He, X.-M. Lin, H. Chan, L. Vukovic, P. Kràl, H.M. Jaeger, Diffusion and filtration properties of self-assembled gold nanocrystal membranes. Nano Lett. **11**(6), 2430–2435 (2011)

84. S.K. Hau, H.-L. Yip, N.S. Baek, J. Zou, K. O'Malley, A.K.-Y. Jen, Air-stable inverted flexible polymer solar cells using zinc oxide nanoparticles as an electron selective layer. Appl. Phys. Lett. **92**(25), 253301 (2008)

85. N. Olichwer, E.W. Leib, A.H. Halfar, A. Petrov, T. Vossmeyer, Cross-linked gold nanoparticles on polyethylene: resistive responses to tensile strain and vapors. ACS Appl. Mater. Interf. **4**(11), 6151–6161 (2012)

86. J. He, X.-M. Lin, R. Divan, H.M. Jaeger, In-situ partial sintering of gold-nanoparticle sheets for SERS applications. Small **7**(24), 3487–3492 (2011)

87. K. Saha, S.S. Agasti, C. Kim, X. Li, V.M. Rotello, Gold nanoparticles in chemical and biological sensing. Chem. Rev. **112**(5), 2739–2779 (2012)

88. A. Tricoli, S.E. Pratsinis, Dispersed nanoelectrode devices. Nat. Nanotechnol. **5**(1), 54–60 (2010)

89. C.-F. Chen, S.-D. Tzeng, H.-Y. Chen, K.-J. Lin, S. Gwo, Tunable plasmonic response from alkanethiolate-stabilized gold nanoparticle superlattices: Evidence of near-field coupling. J. Amer. Chem. Soc. **130**(3), 824–826 (2008)

90. G. Yang, L. Hu, T.D. Keiper, P. Xiong, D.T. Hallinan, Gold nanoparticle monolayers with tunable optical and electrical properties. Langmuir **32**(16), 4022–4033 (2016)

91. S. Chen, R.S. Ingram, M.J. Hostetler, J.J. Pietron, R.W. Murray, T.G. Schaaff, J.T. Khoury, M.M. Alvarez, R.L. Whetten, Gold nanoelectrodes of varied size: transition to molecule-like charging. Science **280**(5372), 2098–2101 (1998)

92. M.-C. Daniel, D. Astruc, Gold nanoparticles: Assembly, supramolecular chemistry, quantum-size-related properties, and applications toward biology, catalysis, and nanotechnology. Chem. Rev. **104**(1), 293–346 (2004)

93. Z. Nie, A. Petukhova, E. Kumacheva, Properties and emerging applications of self-assembled structures made from inorganic nanoparticles. Nat. Nanotechnol. **5**(1), 15–25 (2010)

94. A. Dong, J. Chen, P.M. Vora, J.M. Kikkawa, C.B. Murray, Binary nanocrystal superlattice membranes self-assembled at the liquid–air interface. Nature **466**(7305), 474–477 (2010)

95. K.E. Mueggenburg, X.-M. Lin, R.H. Goldsmith, H.M. Jaeger, Elastic membranes of close-packed nanoparticle arrays. Nat. Mater. **6**(9), 656–660 (2007)

96. X.W. Gu, X. Ye, D.M. Koshy, S. Vachhani, P. Hosemann, A.P. Alivisatos, Tolerance to structural disorder and tunable mechanical behavior in self-assembled superlattices of polymer-grafted nanocrystals. Proc. Natl. Acad. Sci. **114**(11), 2836–2841 (2017)

97. K.M. Salerno, D.S. Bolintineanu, J.M.D. Lane, G.S. Grest, High strength, molecularly thin nanoparticle membranes. Phys. Rev. Lett. **113**(25), 258301 (2014)

98. Y. Wang, J. Liao, S.P. McBride, E. Efrati, X.-M. Lin, H.M. Jaeger, Strong resistance to bending observed for nanoparticle membranes. Nano Lett. **15**(10), 6732–6737 (2015)

99. Y. Wang, P. Kanjanaboos, E. Barry, S. Mcbride, X.-M. Lin, H.M. Jaeger, Fracture and failure of nanoparticle monolayers and multilayers. Nano Lett. **14**(2), 826–830 (2014)

100. J. Hure, B. Roman, J. Bico, Stamping and wrinkling of elastic plates. Phys. Rev. Lett. **109**(5), 054302 (2012)

101. C. Lee, X. Wei, J.W. Kysar, J. Hone, Measurement of the elastic properties and intrinsic strength of monolayer graphene. Science **321**(5887), 385–388 (2008)

102. L.H. Dudte, E. Vouga, T. Tachi, L. Mahadevan, Programming curvature using origami tessellations. Nat. Mater. **15**(5), 583–588 (2016)

103. P. Kanjanaboos, A. Joshi-Imre, X.-M. Lin, H.M. Jaeger, Strain patterning and direct measurement of Poisson's ratio in nanoparticle monolayer sheets. Nano Lett. **11**(6), 2567–2571 (2011)

104. L.A. Girifalco, R.J. Good, A theory for the estimation of surface and interfacial energies. I. Derivation and application to interfacial tension. J. Phys. Chem. **61**(7), 904–909 (1957)

105. S. Wu, Calculation of interfacial tension in polymer systems. J. Polym. Sci., Part C: Polym. Symp. **34**(1), 19–30 (2007)

106. J. He, P. Kanjanaboos, N.L. Frazer, A. Weis, X.-M. Lin, H.M. Jaeger, Fabrication and mechanical properties of large-scale freestanding nanoparticle membranes. Small **6**(13), 1449–1456 (2010)

107. Y. Wang, H. Chan, B. Narayanan, S.P. McBride, S.K.R.S. Sankaranarayanan, X.-M. Lin, H.M. Jaeger, Thermomechanical response of self-assembled nanoparticle membranes. ACS Nano **11**(8), 8026–8033 (2017)

108. S.D. Griesemer, S.S. You, P. Kanjanaboos, M. Calabro, H.M. Jaeger, S.A. Rice, B. Lin, The role of ligands in the mechanical properties of Langmuir nanoparticle films. Soft Matt. **13**(17), 3125–3133 (2017)

109. J.C. Crocker, D.G. Grier, Methods of digital video microscopy for colloidal studies. J. Colloid and Interface Sci. **179**(1), 298–310 (1996)

110. J. Weertman, J. Weertman, *Elementary Dislocation Theory* (Oxford University Press, Oxford, 1992)

111. L. Pocivavsek, R. Dellsy, A. Kern, S. Johnson, B. Lin, K.Y.C. Lee, E. Cerda, Stress and fold localization in thin elastic membranes. Science **320**(5878), 912–916 (2008)

112. P. Kim, M. Abkarian, H.A. Stone, Hierarchical folding of elastic membranes under biaxial compressive stress. Nat. Mater. **10**(12), 952–957 (2011)

113. D. Vella, B. Davidovitch, Regimes of wrinkling in an indented floating elastic sheet. Phys. Rev. E **98**(1), 013003 (2018)

114. C. Androulidakis, K. Zhang, M. Robertson, S. Tawfick, Tailoring the mechanical properties of 2D materials and heterostructures. 2D Mater. **5**(3), 032005 (2018)

115. K. Kang, K.-H. Lee, Y. Han, H. Gao, S. Xie, D.A. Muller, J. Park, Layer-by-layer assembly of two-dimensional materials into wafer-scale heterostructures. Nature **550**(7675), 229–233 (2017)

116. M.Z. Hasan, C.L. Kane, Colloquium: topological insulators. Rev. Mod. Phys. **82**(4), 3045–3067 (2010)

117. M.C. Rechtsman, J.M. Zeuner, Y. Plotnik, Y. Lumer, D. Podolsky, F. Dreisow, S. Nolte, M. Segev, A. Szameit, Photonic floquet topological insulators. Nature **496**(7444), 196–200 (2013)

118. E. Prodan, C. Prodan, Topological phonon modes and their role in dynamic instability of microtubules. Phys. Rev. Lett. **103**(24), 248101 (2009)

119. P. Wang, L. Lu, K. Bertoldi, Topological phononic crystals with one-way elastic edge waves. Phys. Rev. Lett. **115**(10), 104302 (2015)

120. G. Jotzu, M. Messer, R. Desbuquois, M. Lebrat, T. Uehlinger, D. Greif, T. Esslinger, Experimental realization of the topological Haldane model with ultracold fermions. Nature **515**(7526), 237 (2014)

121. M. Aidelsburger, M. Lohse, C. Schweizer, M. Atala, J.T. Barreiro, S. Nascimbène, N.R. Cooper, I. Bloch, N. Goldman, Measuring the Chern number of Hofstadter bands with ultracold bosonic atoms. Nat. Phys. **11**(2), 162 (2015)

122. F.D.M. Haldane, Model for a quantum Hall effect without Landau levels: condensed-matter realization of the "Parity Anomaly". Phys. Rev. Lett. **61**(18), 2015–2018 (1988)

123. O.R. Bilal, A. Foehr, C. Daraio, Bistable metamaterial for switching and cascading elastic vibrations. Proc. Natl. Acad. Sci. **114**(18), 4603–4606 (2017)

124. F.D.M. Haldane, S. Raghu, Possible realization of directional optical waveguides in photonic crystals with broken time-reversal symmetry. Phys. Rev. Lett. **100**(1), 013904 (2008)

125. J. Ningyuan, C. Owens, A. Sommer, D. Schuster, J. Simon, Time- and site-resolved dynamics in a topological circuit. Phys. Rev. X **5**(2), 021031 (2015)

126. V. Peano, C. Brendel, M. Schmidt, F. Marquardt, Topological phases of sound and light. Phys. Rev. X **5**(3), 031011 (2015)

127. Z. Wang, Y.D. Chong, J.D. Joannopoulos, M. Soljacic, Reflection-free one-way edge modes in a gyromagnetic photonic crystal. Phys. Rev. Lett. **100**(1), 013905 (2008)

128. R. Fleury, A.B. Khanikaev, A. Alù, Floquet topological insulators for sound. Nat. Commun. **7**, 11744 (2016)

129. A.B. Khanikaev, R. Fleury, S.H. Mousavi, A. Alù, Topologically robust sound propagation in an angular-momentum-biased graphene-like resonator lattice. Nat. Commun. **6**, 8260 (2015)

130. A. Souslov, B.C. van Zuiden, D. Bartolo, V. Vitelli, Topological sound in active-liquid metamaterials. Nat. Phys. **13**(11), 1091–1094 (2017)

131. S. Maayani, R. Dahan, Y. Kligerman, E. Moses, A.U. Hassan, H. Jing, F. Nori, D.N. Christodoulides, T. Carmon, Flying couplers above spinning resonators generate irreversible refraction. Nature **558**(7711), 569–572 (2018)

132. J.E. Avron, R. Seiler, B. Simon, Homotopy and quantization in condensed matter physics. Phys. Rev. Lett. **51**(1), 51–53 (1983)

133. R.B. Laughlin, Quantized Hall conductivity in two dimensions. Phys. Rev. B **23**(10), 5632–5633 (1981)

134. J.L. Mañes, F. Guinea, M.A.H. Vozmediano, Existence and topological stability of Fermi points in multilayered graphene. Phys. Rev. B **75**(15), 155424 (2007)

135. N.W. Ashcroft, N.D. Mermin, *Solid State Physics*, 1st edn. (Cengage Learning, New York, 1976)

136. C.L. Kane, E.J. Mele, Quantum spin Hall effect in graphene. Phys. Rev. Lett. **95**(22), 226801 (2005)

137. D.M. Sussman, O. Stenull, T.C. Lubensky, Topological boundary modes in jammed matter. Soft Matt. **12**(28), 6079–6087 (2016)

138. P.M. Chaikin, T.C. Lubensky, *Principles of Condensed Matter Physics* (Cambridge University Press, Cambridge, 2000)

139. Z. Yang, F. Gao, X. Shi, X. Lin, Z. Gao, Y. Chong, B. Zhang, Topological acoustics. Phys. Rev. Lett. **114**(11), 114301 (2015)

140. A.S. Meeussen, J. Paulose, V. Vitelli, Geared topological metamaterials with tunable mechanical stability. Phys. Rev. X **6**(4), 041029 (2016)

141. D.J. Thouless, Wannier functions for magnetic sub-bands. J. Phys. C: Solid State Phys. **17**(12), L325 (1984)

142. Y. Huo, R.N. Bhatt, Current carrying states in the lowest Landau level. Phys. Rev. Lett. **68**(9), 1375–1378 (1992)

143. T. Thonhauser, D. Vanderbilt, Insulator/Chern-insulator transition in the Haldane model. Phys. Rev. B **74**(23), 235111 (2006)

144. M. Florescu, S. Torquato, P.J. Steinhardt, Designer disordered materials with large, complete photonic band gaps. Proc. Natl. Acad. Sci. **106**(49), 20658–20663 (2009)

145. D. Weaire, M.F. Thorpe, Electronic properties of an amorphous solid. I. A simple tight-binding theory. Phys. Rev. B **4**(8), 2508–2520 (1971)

146. D. Weaire, Existence of a gap in the electronic density of states of a tetrahedrally bonded solid of arbitrary structure. Phys. Rev. Lett. **26**(25), 1541–1543 (1971)

147. R. Haydock, V. Heine, M.J. Kelly, Electronic structure based on the local atomic environment for tight-binding bands. J. Phys. C: Solid State Phys. **5**(20), 2845 (1972)

148. A. Kitaev, Anyons in an exactly solved model and beyond. Ann. Phys. **321**(1), 2–111 (2006)

149. E. Prodan, Non-commutative tools for topological insulators. New J. Phys. **12**(6), 065003 (2010)

150. R. Bianco, R. Resta, Mapping topological order in coordinate space. Phys. Rev. B **84**(24), 241106 (2011)

151. B.I. Halperin, Quantized Hall conductance, current-carrying edge states, and the existence of extended states in a two-dimensional disordered potential. Phys. Rev. B **25**(4), 2185–2190 (1982)

152. L. Fu, C.L. Kane, E.J. Mele, Topological insulators in three dimensions. Phys. Rev. Lett. **98**(10), 106803 (2007)

153. E. Prodan, H. Schulz-Baldes, in *Bulk and Boundary Invariants for Complex Topological Insulators*. Mathematical Physics Studies (Springer, Cham, 2016)

154. C. Miehe, F. Welschinger, M. Hofacker, Thermodynamically consistent phase-field models of fracture: variational principles and multi-field FE implementations. Int. J. Numer. Methods Eng. **83**(10), 1273–1311 (2010)

155. A.J. Rosakis, O. Samudrala, D. Coker, Cracks faster than the shear wave speed. Science **284**(5418), 1337–1340 (1999)

156. P.J. Petersan, R.D. Deegan, M. Marder, H.L. Swinney, Cracks in rubber under tension exceed the shear wave speed. Phys. Rev. Lett. **93**(1), 015504 (2004)

157. D.J. Struik, *Lectures on Classical Differential Geometry*, 2nd edn. (Dover Publications, New York, 1988)

CPSIA information can be obtained
at www.ICGtesting.com
Printed in the USA
LVHW080351100220
646387LV00003B/217